The Cambridge Manuals of Science and
Literature

HOUSE-FLIES AND HOW THEY
SPREAD DISEASE

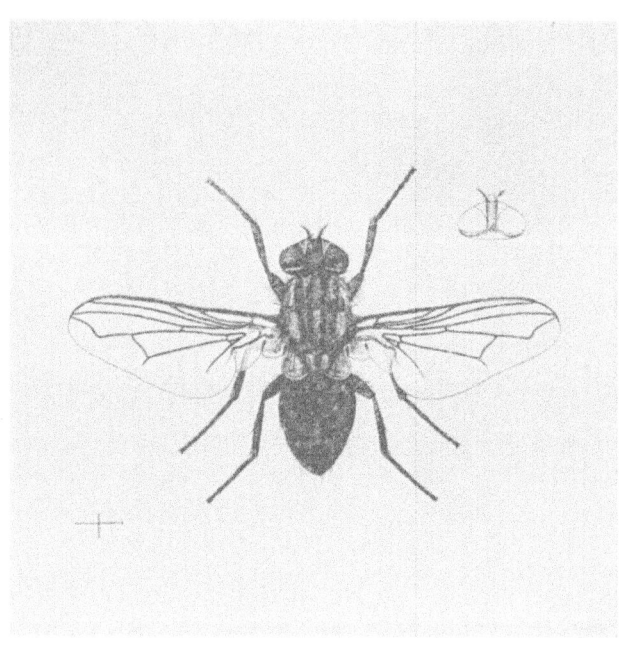

Frontispiece. The House-fly, *Musca domestica*. Female.

HOUSE-FLIES
AND HOW THEY
SPREAD DISEASE

BY

C. G. HEWITT, D.Sc.

Dominion Entomologist,
Ottawa, Canada

Cambridge:
at the University Press
1914

CAMBRIDGE UNIVERSITY PRESS
Cambridge, New York, Melbourne, Madrid, Cape Town,
Singapore, São Paulo, Delhi, Tokyo, Mexico City

Cambridge University Press
The Edinburgh Building, Cambridge CB2 8RU, UK

Published in the United States of America by Cambridge University Press, New York

www.cambridge.org
Information on this title: www.cambridge.org/9781107673052

© Cambridge University Press 1912

First published 1912
Reprinted 1914
First paperback edition 2011

A catalogue record for this publication is available from the British library

ISBN 978-1-107-67305-2 Paperback

*With the exception of the coat of arms
at the foot, the design on the title page is a
reproduction of one used by the earliest known
Cambridge printer, John Siberch, 1521*

PREFACE

ABOUT eight years ago, on being asked for some information of a special kind regarding the house-fly, I was surprised to find after looking into the matter, that our knowledge of this insect was of the most meagre character. Notwithstanding the fact that it is usually the first animal with which man makes his acquaintance on entering the world and is certainly his most constant companion through life, little attention has been paid to this the commonest of insects. A few studies of its life-history and development had been made in the United States and Germany but there was no accurate information concerning its structure, and the general ignorance of its biology and habits was astonishing, to say the least.

In view of this hiatus in our knowledge and the increasing necessity for such information on account of the accumulating evidence as to the disease carrying character of the house-fly, I commenced a study of its structure, development, and biology with especial reference to its relation to the dissemination of disease. The results of the greater portion of these studies were published in *The Quarterly Journal of Microscopical Science* in 1907, 1908 and 1909 respectively. During the last two or three years the appreciation of the true nature of the fly in its relation to man has been responsible for the carrying out by numerous workers of a considerable number of investigations on the relation of the house-fly to disease, etc.

In this contribution to *The Cambridge Manuals of Science and Literature* it has been my endeavour to avoid, so far as is possible, the use of technical terms unfamiliar to the lay mind and the inclusion of matter which is of interest chiefly to the specialist.

I am indebted to Mr H. T. Güssow, Dominion Botanist, Ottawa, for the photograph of *Empusa* (fig. 13) and to Mr C. T. Brues of Harvard University for the original photograph of fig. 5. I am responsible for the rest of the illustrations.

Through the investigations and educational work of a comparatively few workers, there has been during the last few years an awakening in the mind of the general public, so resolutely indifferent to such matters, of an interest in the fact that the house-fly is something more than an irritant to elderly gentle-men and an object of interest to babies. It is now more generally realised that it is a serious menace to the health of the community. This little volume has

been written in the hope that it will not only bring
home to a greater number of people the true nature
of the house-fly, but also indicate the means to be
taken to eradicate it or render it no longer a menace.
It is the scientist's mission to discover the path, but
he cannot, after having blazed the trail, make the
public follow therein, even though the path be clear
and they see the beautiful country beyond. I have
frequently been told by people with good intentions
that they can do nothing. They can. They plead
the indifference or absolute opposition of local
authorities to sanitary improvements. Then, I say,
elect only those who are pledged to make sanitary
changes which will give the children a fair chance
and the people healthy surroundings. It can be done
and has been done whenever and wherever the people
so determine. The educational work necessary is not
easy; it is often discouraging. Early in my work
the editor of a well-known London weekly journal
recommended my incarceration in a lunatic asylum,
and another eminent medical man suggested that
had I propounded such doctrines a few years ago
a commission might have been appointed to inquire
into the state of my mind. But it is ever so, and that
stage in the history of this doctrine is past. The
hostile period is practically over ; the indifferent
and apathetic period is waning. People can avoid
hypotheses but they cannot escape facts. We have

arrived at the stage where the facts are incontrovertible.

Therefore, let there be a more widespread determination on the part of the people generally to take the steps or to insist on measures being taken, which our present knowledge of the subject indicates as being necessary, to eradicate, so far as is humanly possible, this potential disease carrier and constant frequenter of filth. Such action will most surely result in a vast improvement in the sanitary conditions of our cities and towns and, by the decrease of intestinal disease, in the health and welfare of the people generally.

C. G. H.

OTTAWA, CANADA,
April 1912.

CONTENTS

LIST OF ILLUSTRATIONS

PART I

CHAPTER I

INTRODUCTION

NOT the least of the striking features of recent scientific progress has been the discovery of the vital relationship of insect life to humanity. Daily are we realising with increasing knowledge the power and majesty of Beelzebub[1]. Through all ages mosquitoes have inflicted greater destruction than all the armies of the world. Within the last few years we have witnessed the sacrifice of thousands of lives to the Tse-tse fly's taste for blood and the kindred of the irritating flea have been shown to play no mean *rôle*

---

[1] Beelzebub. *Baal*, meaning 'lord' or 'god' and *zebub*, usually identified with the Hebrew word of the same spelling, meaning 'flies.' It is interesting to note that the alternative spelling is Beelzebul; *zebul* is for *zebel* sometimes used for 'dung.' From what is now known of the breeding habits of flies these alternatives would appear to have a closer and more apparent relationship!

in the spread of one of our most deadly diseases. The power of transmitting disease germs however is not confined to insects of blood-sucking habits. Step by step evidence, both circumstantial and exact, has been accumulated until it has now been shown and is an undisputed fact that where the necessary conditions occur, our most common and widespread insect the house-fly is a prominent factor in the carriage of infection. Living with us in our homes, feeding off our tables, we have an insect whose befouled body and limbs may at any time be bearing micro-organisms of a dangerous nature. It does not seem so long ago since we were being taught to regard this insect as a very respectable member of our household. *Tempora mutantur.* No longer will children be taught that it is wrong to kill a fly but rather the first step to be taken by one who would seek to mould himself after the pattern of St George.

The discovery of the serious *rôle* which the house-fly may play under certain conditions, was hardly less astonishing than the revelation of our profound ignorance of its life-history, habits and bionomics generally. In this respect it afforded another example of a not infrequent occurrence in biology, namely, the passing over of the commonplace in the search for the unique and rare. Zoological literature is replete with minute descriptions of creatures which few have seen, or ever will see, and which have not the remotest relationship to man. On the other hand here we have an insect, undoubtedly the commonest and most widely spread of all insects, which has accompanied man wherever he has travelled whether it be into the torrid heat of the tropics or into the icy regions of the north, which is his companion from the time he enters the world until he leaves it, which amuses the young and annoys the old, but notwithstanding all this, it has been too common a creature to be deemed worthy of serious study. With a single exception no attempt has been made until within the last few years to make a careful study of the house-fly. In 1790[1] Wilhelm von Gleichen published in Nürnberg an excellent account of the life-history and habits of this insect which was illustrated by some exceptionally interesting plates, and when the present writer commenced a detailed study of the house-fly in 1905 Gleichen's work was still the most exhaustive account available. The older naturalists from Reaumur (1738) onwards included short accounts of the house-fly in their general works and during the last century fragmentary observations were recorded. The last complete account of the life-history had been written by Packard in 1874, which together with some additional investigations of Howard in 1898 and 1902, constituted practically our entire knowledge of the

[1] This is the date of the edition in my possession but an earlier edition would appear to have been published in 1764 or 1766.

commonest insect. Never was the adage 'Familiarity
breeds contempt' more forcibly illustrated. The
placing of its human relationship on a different basis
was responsible for a kindling of interest in and a
desire for knowledge of the house-fly with the result
that it was in danger of becoming almost a cult.
Fortunately this did not happen, but instead a considerable number of investigations bearing on its
bionomics and relations to disease were carried out
and are still being prosecuted both in England and
in the United States. Its relation to public health
was deemed sufficiently serious to warrant an inquiry
by the Local Government Board: this inquiry is now
being carried on and already several valuable reports
have been issued. In the United States a very active
campaign is being waged on all sides against the
house-fly as it is recognised to be a serious factor in
the transmission of zymotic diseases and as being
synonymous with insanitary conditions. No small
credit for this activity is due to the primary and continued efforts of Dr L. O. Howard, the Entomologist
of the United States Department of Agriculture. As
illustrating the popular feeling with regard to fly
campaign in the United States it may be mentioned
that the Mayor of the capital of one of the States was
elected almost solely on the strong stand which he
had taken in advocating anti-fly measures. This
sudden change of opinion, which has already affected

and is reflected in the bye-laws relating to public
health matters, is of more than ordinary interest
and is fully in keeping with the spirit of the age.
One is reminded of the description of the fly which
Ruskin gives us. When Menelaus the King of Sparta
invokes the goddess Athena for strength to withstand
Hector, she gives him the courage of the most fearless
and audacious of creatures, namely, the fly. 'The
common house-fly,' Ruskin says, 'is the most perfectly
free and republican of creatures. There is no courtesy
in him; he does not care whether it is a king or clown
whom he teases and in every step of his swift,
mechanical march and in every pause of his resolute
observation, there is one and the same perfect expres-
sion of perfect egotism, perfect independence and
self-confidence and conviction of the world having
been made for flies. Your fly free in the air, free in
the chamber, a black incarnation of caprice, wandering,
investigating, fleeting, flitting, feasting at his will with
rich variety of feast, from the heaped sweets in the
grocer's window to those of the butcher's back-yard
and from the galled place on your horse's neck to the
brown spot on the road from which, as the hoof
disturbs him, he rises with angry republican buzz;
what freedom is like his?'

In contemplating the change of public opinion
towards the fly and our present attitude towards it
which our investigations have brought about we are

made increasingly conscious of the truth of Gilbert White's sage words written in 1777, that :

'The most insignificant insects and reptiles are of much more consequence and have much more influence in the economy of nature than the incurious are aware of.'

CHAPTER II

THE STRUCTURE OF THE FLY

In order to understand how a fly lives, moves and has its being it is necessary to know something of the manner in which its body is built up, in other words to understand its structure. This is no dry aggregation of unexplicable details but a story of how a creature has been perfected in accordance with the requirements of its life and habits.

The typical insect has two pairs of wings such as we find in the butterfly, the grasshopper, or the beetle. The house-fly, however, has a single pair only and on this account it is included with all other flies in a large family known as the *Diptera* or two-winged flies. The hind pair of wings which are the undeveloped ones, are represented by a small pair of drumstick-shaped appendages known as 'balancers' owing to the fact that they are considered by some to

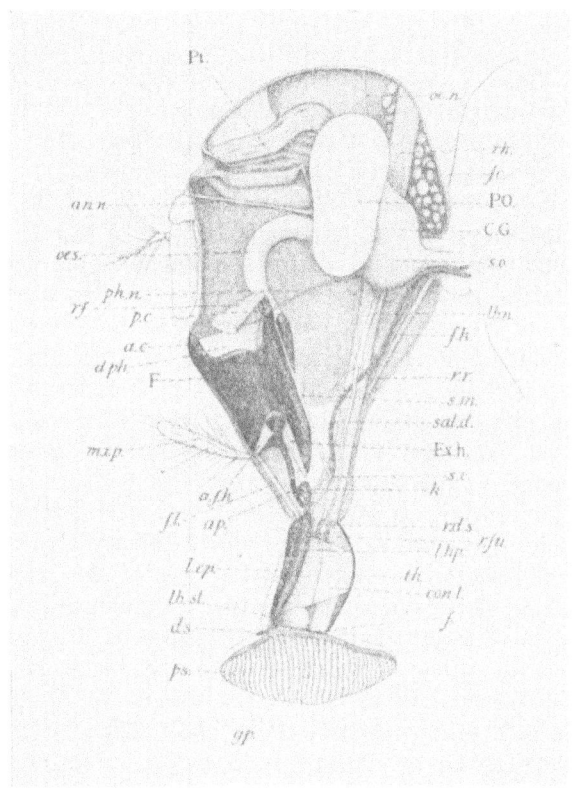

Fig. 1.   Interior of the head of house-fly.   Left side of head and air-
sacs removed.   Of the various parts the following may be noted :
   Pt., eversible frontal sac; *oes.*, oesophagus; *F.*, pharyngeal suction pump; *d.ph.*, muscles working same; *P.O.*, *C.G.*, *s.o.*, brain;
sal.d., salivary duct with its valve *s.v.*; *lb.st.*, salivary gland of
oral lobe; *p.s.*, channel or pseudotrachea of oral lobe; *l.ep.*, upper
lip; *l.hp.*, lower lip or floor of mouth; *oc.n.*, *an.n.*, *ph.n.*, *lb.n.*,
nerves.   The rest of the structures are chiefly muscles for the
retraction and movements of the proboscis, etc.

have some relation to the maintenance of equilibrium. The characteristic divisions of the body are well marked.

The whole structure of the fly is admirably adapted to a flitting, aerial life. A pair of large compound eyes occupies almost the whole of the hemispherical head and provides the insect with a wide field of vision. The greater portion of the interior of the middle-body or thorax is occupied with muscles used in flying. In the abdomen a pair of large airsacs gives great buoyancy to the body of the insect, which buoyancy is increased further by air-sacs situated in the thorax and head (fig. 3). The provision of these air-sacs together with a well developed tracheal system, as the system of silvery air-tubes permeating all the organs and tissues of the body is called, has an important relation to the remarkable degree of activity evinced by these insects. In insects these air-tubes or tracheae take the place of the blood-vessels and capillaries found in higher animals, and as in birds a rich blood supply is associated with an active life so likewise in an active insect as the house-fly we find a rich tracheal supply.

The greater portion of the hemispherical head of the fly is occupied by a pair of large compound eyes; each being composed of about four thousand faceted individual eyes which together apparently produce a single somewhat blurred image and not thousands of

Fig. 2. Longitudinal section of alimentary canal of house-fly. *ph.*, pharyngeal suction pump; *oe.*, oesophagus; *pt.*, frontal sac; *c.g.*, brain; *s.d.*, salivary duct; *P.V.*, proventriculus; *V.*, chyle stomach or ventriculus; *C.*, crop or sucking stomach; *p.i.*, proximal intestine; *R.*, rectum; *r.g.*, rectal glands.

separate images. On the top of the head in the space
between the eyes are three simple eyes arranged in a
triangle. The width of this space on the top of the
head between the eyes serves as a means of distinguishing the male from the female which is otherwise
impossible without a more careful examination. In
the male the eyes are separated by about one-fifth
the breadth of the head, but in the female this space
is wider, being about one-third the breadth of the
head.

The proboscis or, as it is sometimes inaccurately
called, 'tongue' of the fly, represents the enormously
modified mouth parts or jaws of other insects such as
beetles and grasshoppers which chew their food. The
mouth parts of the house-fly have been adapted to a
sucking function not combined with or involving a
previous piercing function, as in the case of the
mosquito and flea which pierce before sucking. The
house-fly cannot pierce the softest skin, as the parts
which are used to pierce the skin in the case of the
stable-fly (*Stomoxys calcitrans*) are reduced to harm-
less proportions in the house-fly, and the proboscis is
terminated by a pair of soft cushion-like lobes or lips,
called the oral lobes, which form together a heart-
shaped structure having the aperture leading into the
mouth situated in the middle. The popular idea that
the house-fly 'bites' at certain times is incorrect, as
it cannot bite; the idea is due to a confusion of

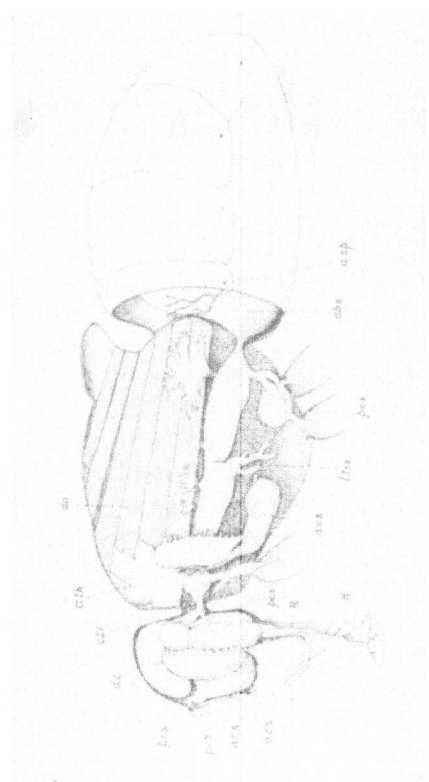

Fig. 3. Chief portion of respiratory or tracheal system of house-fly. The air-sacs supplied by the posterior thoracic spiracle are omitted. *d.c, p.c.s., p.op., a.c.s., v.c.s., p.c.s.,* air-sacs of head; *R.,* rostrum; and *H.,* haustellum together forming the proboscis; *a.th.,* anterior thoracic spiracle; *a.v.s., l.tr.s., p.v.s.,* air-sacs in thorax; *d.o,* large muscles of thorax concerned in flight; *a.b.s.,* one of abdominal air-sacs; *a.sp.,* abdominal spiracles.

the house-fly with its cousin the stable-fly, to which it has a general resemblance in the eyes of the unskilled observer.

The structure of the proboscis and its mechanism are of unusual interest (see fig. 1). Traversing the inner or under sides of each of the cushion-like oral lobes, which when not in use are closed like the valves of a mussel, are over thirty small open channels which are called pseudo-tracheae (*ps.*) from their similarity to the annulated tracheal tubes. The rings however are not complete, being open on one side to form an almost cylindrical groove. All these channels run into main channels situated on the inner edges of the oral lobes and the main channels open into the mouth. The mouth leads by way of a grooved channel into a powerful muscular pump, the pharyngeal pump (see *ph.* fig. 2), into which the liquid food is pumped up. The fly has two pairs of salivary glands. A large pair of coiled white tubes lie in the thorax; the ducts from these join together in the head and open into the mouth. At the base of the oral lobes is another pair of spherical glands (*lb. st.*) which open on to the oral surface and probably keep this surface in a moist condition. In a state of rest the proboscis is carried bent up by means of special muscles in an elbow fashion in the inside of the lower portion of the head with the oral lobes closed upon each other. When the fly alights on food the

proboscis is protruded by the combined action of the
air-sacs in the head and the blood. The blood in the
head is not confined to blood-vessels but occupies the
cavity of the head not taken up by the capacious air-
sacs, brain and other organs. Consequently, if there
is an expansion of the air-sacs in the head the blood
in the head is forced into the protrusible proboscis
which is thereby extended, as one may extend the
drawn-in fingers of a glove by blowing into it.
It was formerly thought that the proboscis was
extended by means of air but my investigations
showed that the proboscis did not contain air-sacs
but definite tracheae, incapable of expansion. The
surface of each of the oral lobes is provided with a
large number of sense organs and many of the long
hairs fringing and on the back of the oral lobes are
sensory in character. These sense organs probably
enable the fly to taste and smell. From the pharyngeal
pump the oesophagus passes through the cerebral
ganglia or 'brain' of the fly into the thorax by way
of the narrow neck (see fig. 2). After entering the
thorax it opens into the proventriculus but before
doing so gives off a duct on the under side which
leads into the crop (c.) situated at the front end of
the abdomen and on the ventral side. The crop is a
thin walled vesicle and its function and relation to
the ventriculus will be explained in the next chapter
in describing the fly's method of feeding. The

proventriculus (PV) is shaped like a small flattened sphere and its interior is almost filled by a circular flattened and perforated plug which appears to have a valvular function, in fact the function of the proventriculus of the fly would seem to be that of a combined pump and valve operated at will by the fly.

The proventriculus opens into the chyle-stomach or ventriculus (V). This traverses the thorax and joins the intestine in the front end of the abdomen. Its walls are thrown into a number of folds which are subdivided forming a number of little chambers or sacs which contain large digestive cells. The intestine is thrown into a number of coils in the abdomen and is divided into two distinct parts by the junction with it of a pair of long coiled tubes, the malpighian tubes, of a yellowish-white colour. These tubes have an excretory function and may correspond roughly to the kidneys of higher animals. At the hind end of the abdomen the intestine opens into the rectum (R) which opens externally by the anus. The rectum is swollen in the middle to form a cavity. This rectal cavity contains two pairs of peculiar conical rectal glands ($r.gl.$) which are richly supplied with tracheae, and probably extract waste substances from the blood to excrete them in the rectum.

The respiratory system or tracheal system of the house-fly (fig. 3) is very highly developed as we have already seen. Altogether it occupies more space in

the body of the fly than any other set of organs. It
consists of three parts; the spiracles or breathing
pores situated on the sides of the body, air-sacs and
air-tubes or tracheae. A large pair of spiracles (*a.th.*)
is situated over the bases of the first pair of legs.
These spiracles supply air to the following: a set of
air-sacs which fill up the vacant space in the head; a
series of air-sacs in the thorax which give off tracheae
to the muscles and legs, and two large air-sacs which
occupy, in some cases, almost the whole of the front end
of the abdomen and give off tracheae to the viscera.
Above and behind the bases of the last pair of legs is
another pair of spiracles which supply air to the large
muscles of the thorax in that region. In addition to
these thoracic spiracles there are a number of pairs
of spiracles situated at the sides of the abdominal
segments. In the male there are seven pairs of abdominal spiracles but in the female I have only been
able to discover five pairs. All these spiracles com-
municate with tracheae which ramify among the
intestinal organs of the abdomen.

The blood system of the fly is simple. The body
cavity forms a blood cavity so that all the organs and
muscles etc., are bathed by the blood fluid which is
colourless and contains fatty corpuscles. There is a
muscular tube or 'heart' lying in a cavity immediately
under the dorsal side of the abdomen. It extends
from the posterior end to the anterior end of the

abdomen and is divided into four chambers, each having a pair of openings into which the blood is sucked, so to speak, from the pericardial cavity. The heart is continued as a dorsal vessel along the chyle stomach and appears to terminate in a mass of cells behind the proventriculus. Associated with the blood system is a diffuse structure known as the fat body which consists of a large number of very large fat cells well supplied with tracheae. The size of the fat body varies considerably; just before hibernation it seems to fill almost the whole of the abdominal cavity and after hibernation it is found to have shrunk almost to nothing. While it may have some excretory function, it would appear that it also stores up the products of digestion which it obtains from the blood with which it is bathed.

If a female fly is dissected during the summer months its abdomen will be found practically filled with white cylindrical eggs, packed together like cigars in two large bundles (fig. 4). Each of these bundles which are the enlarged ovaries contains about seventy strings of eggs in various stages of development, and the ovaries open into two ducts which join together to form a central duct opening into the telescopic ovipositor (*ovp.*). Connected with this central oviduct are certain glands and a set of three small black vesicles (*sp.*) which store the spermatozoa received from the male during coitus.

The long telescopic ovipositor is composed of the four last segments of the abdomen which can be retracted entirely within the abdomen. When the fly lays its eggs the ovipositor is extended and when fully extended it is as long as the abdomen. The ovipositor of the female can be extended by gently

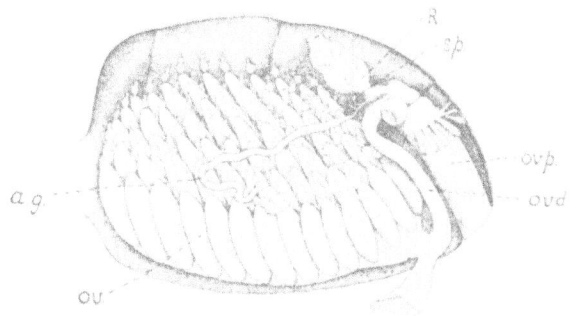

Fig. 4. Longitudinal section of abdomen of female house-fly. Left half removed together with intestine and air-sacs, etc.

    a.g., accessory gland; *ov.*, ovary containing fully formed and undeveloped eggs; *ovd.*, oviduct; *ovp.*, telescopic ovipositor withdrawn into abdomen; *sp.*, spermatheca; *R.*, rectum.

compressing the abdomen with the fingers. The possession of an extensile ovipositor is of great importance as the fly is thereby enabled to deposit its eggs in the crevices of the substance chosen as the nidus for the larvae. The eggs are thus kept suitably moist, and the young larvae, which shun the light, can

begin to feed almost immediately. The internal
reproductive organs of the male consist of a pair of
small brown pear-shaped testes which open by fine
ducts into a common ejaculatory duct. The external
organs of the male consist of a chitinous penis and
accessory plates and it is interesting to note that the
terminal segments of the abdomen which form these
genital plates are asymmetrical in character.

CHAPTER III

LIFE-HISTORY AND BREEDING HABITS OF THE
HOUSE-FLY

THE house-fly, like the butterfly and the beetle,
undergoes a complete metamorphosis in its development from the egg to the perfect insect. That is, it
has a definite larval stage which is followed by a pupal
or resting stage from which the perfectly developed
fly emerges. The larvae are popularly known as
'maggots.' It is in the larval stage that the whole of
the growth takes place and the entire life of the larva
is devoted to the building up of tissues from which
will be derived the tissues of the adult. There is still
a popular idea that house-flies grow, a fallacy due to
the confusion of this species, *Musca domestica*, with
the Lesser House-fly, *Fannia canicularis*. The
latter species is smaller in size and in consequence is

often wrongly considered to be a 'young' house-fly. The perfect insect is incapable of growth; all the growth is in the larval or maggot stage and when the perfect insect crawls out of the puparium, as will be described later, it is incapable of further growth in size.

One of the most remarkable facts to be observed in connection with the house-fly is its enormous fecundity. This is to be attributed to two chief causes: the universal prevalence of substances that are eminently suitable for the nutrition of the larvae and in which they develop, and the rapidity of the development.

Broadly speaking, house-flies will breed in almost any decaying matter of a vegetable or animal nature and in excrementous substances, if certain conditions, namely those of temperature and moisture, are favourable. Temperature, moisture and the character of the food are the chief factors governing the rate of development. In most substances in which the larvae are able to develop successfully the process of fermentation will be found to be taking place; this process is related to, and a result of, a combination of the previous factors and was recognised by De Geer as early as 1776 as an important condition.

Horse manure forms the chief substance in which the larvae are found and heaps of stable manure are the principal breeding places of these insects. In addition to this they have been reared from, or the larvae have been found feeding upon, the following substances: human excrement, both as isolated faeces and in latrines and mixed with ashes; cow dung; fowl dung; cess-pools; straw and textile fabrics such as woollen or cotton garments and sacking which have been fouled with animal excrement; mushrooms; decaying vegetables, fruits and food stuffs such as potato skins, melon, bananas, pears, apricots, cherries, plums, peaches; bread and milk, boiled egg, bad meat, and rotting grain, such as wheat. They have also been found in spittoons. From these facts it will be understood that wherever there are collections of manure, or excrement, or of waste animal or vegetable substances collectively known as garbage, house-flies are able to breed on the occurrence of a suitable temperature.

Flies normally begin to breed in June and July and continue to October, the greatest activity in this respect being in the hot months of August and September. It has been found, however, that when the temperature is sufficiently high, as sometimes occurs in warm cellars and kitchens, flies are able to deposit their eggs and the larvae will develop during the winter months. A single female fly is capable of depositing from 100 to 150 eggs at one time, and five or six and possibly more batches of eggs can be deposited during its short life. It has been calculated

that if the progeny of a single pair of flies, assuming
that they all lived, were pressed together at the end
of the summer, they would occupy a space of about a
quarter of a million cubic feet.

Fig. 5. Mass of eggs of house-fly.
Photo by C. T. Brues.

The eggs of the fly are cylindrically oval in shape
(fig. 5); one end being slightly broader than the other,
towards which the egg tapers off slightly. They are

pearly white in colour and have two rib-like thick-
enings along one side. The average length is 1 mm.
($\frac{1}{25}$ in.). The fly deposits the eggs usually in groups
in the dark crevices of the substances chosen for the
larval nidus, and the female is enabled to place the
eggs in such favourable positions, where they cannot
undergo desiccation, by means of her lengthy ovipositor.
Flies will rarely penetrate a perfectly dark chamber
to deposit their eggs. In warm weather the larvae
hatch out from eight to twenty-four hours after the
eggs are laid; the young maggot emerging through
a split which is formed along the egg. Before
becoming full grown the larva or maggot casts its
skin twice ; that is, there are three larval stages. The
first moult may take place from twenty to thirty-six
hours after hatching, and the second moult twenty-
four hours later. The length of time spent in the
larval stages, and in fact in all the stages of the life-
history, is dependent on the factors already mentioned.
Temperature is the chief of these factors and with a
high temperature (e.g. 30° to 35° C.) the larvae may
become full grown in five days, the third larval stage
lasting three days.

The full-grown larva (fig. 6) is a creamy white
legless maggot measuring 12 mm. ($\frac{1}{2}$ in.) in length.
The body gradually tapers off from the middle to the
anterior end where it terminates in a pair of oral
lobes. Each oral lobe bears two small sensory

tubercles which appear to assist the larva in perceiving
the difference between light and darkness. The head
of the larva, strictly speaking, is withdrawn into the
anterior end, so that externally the larva is headless.
The posterior half of the body is thicker and terminates bluntly. Twelve distinct segments may be
recognised; in reality there are thirteen, the second
segment being of a double nature. The mouth opens
on the under side of and between the oral lobes which

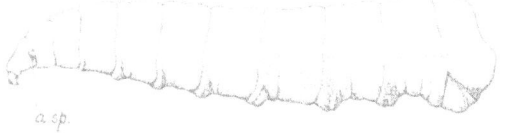

Fig. 6. Larva of house-fly.
a.sp., anterior spiracle or breathing pore.

are readily withdrawn into the succeeding segment.
A black hook-shaped tooth-like process is seen projecting from between the oral lobes. This is part of
the skeleton of the larval head and is used in locomo-
tion and also in tearing up the food which is absorbed
in a semi-liquid form, the solid portions, such as small
pieces of straw, etc., not being taken into the mouth.
At the sides of the second segment is seen a pair of
golden fan-shaped organs (*a.sp.*), each having six to
eight lobes or rays. These are the anterior spiracles

through which air is taken into the respiratory tubes of the larva. A second pair of eye-like spiracles, the posterior spiracles, is found in the middle of the obliquely blunt posterior end of the larva. Each of these consists of a black chitinous ring enclosing three sinuous slits through which the air passes into small chambers at the ends of the pair of thick longitudinal respiratory tubes. On the ventral side of the larva at the anterior edge of each of the sixth to twelfth body segments is a crescentic-shaped pad covered with short recurved spines. These locomotory pads take the place of legs and are used by the larva in conjunction with the mouth hook in travelling backwards and forwards, as one may readily observe if a maggot is watched. The anus is situated between two prominent lobes on the ventral side of the terminal segment. The larva is covered by a thin cuticular integument through which the internal organs may be observed in younger larvae. As the larva becomes mature, the growth of the fat tissues gives it a creamy appearance and the internal organs are obscured.

During its life the larva shuns the light and is affected by changes in the humidity of its food. The drying of the food will cause the larvae to seek the deeper and moister regions and if this is not possible the development is considerably retarded. Retardation in development caused either by the drying up

or insufficiency of food usually results in the production of flies below the normal size.

When full grown the mature larva usually leaves
the moist situation in which it has developed for one
of a drier nature, often crawling for several yards in
search of some dry and sheltered crevice. Here it
rests for a short time preparatory to changing into
the pupal stage. At the beginning of this change a
contraction of the body takes place and the anterior
segments are withdrawn. In this manner it assumes
a cylindrical shape, the anterior and posterior ends
being evenly rounded. In a few hours this barrel-
shaped pupa changes from its original creamy yellow
colour to a rich dark brown. At the posterior end
the two black button-like prominences are the larval
spiracles, and the locomotory pads can still be recognised as roughened areas on the ventral side of the
pupa. The larval skin forms the pupal case or
puparium (fig. 7) in which the larval organs undergo
disintegration and the organs of the future fly are
built up from small groups of cells called imaginal
bodies or discs. From a footless, headless and light-
shunning maggot living in the worst of filth is developed a vivacious, sun-loving aerial insect, an
unbidden guest at the rich man's table; truly one of
the greatest transformations in the world. If the
temperature is high the perfect fly emerges between
the third and fourth day after pupation. In order to

emerge the fly pushes off the anterior end of the
pupal case in two parts, as shown in the figure, by
means of an inflated sac termed the ptilinum on the

Fig. 7. Puparium of House-fly as it appears after emergence of fly.
    a.sp., anterior spiracle of larva; *l.tr.*, moulted breathing tube
or trachea of larva; *n.sp.*, breathing pore of fly while in nymph
stage; *p.sp.*, posterior spiracle of larva.

front of the head. This frontal sac like the horny
tip to the bill of a chick, is only of use in the act of
emerging, and making its way to the air if buried in

the soil. After the fly has emerged from the pupal case, the frontal sac is withdrawn into the head as shown in figures 1 and 2, and its presence is indicated externally by a crescentic slit known as the lunule situated above the bases of the antennae. In a few hours after emerging the wings, which at the time of emergence are loose crumpled sac-like organs, assume their final state and the chitin of the body hardens.

Flies were found to become sexually mature in ten to fourteen days after emergence from the pupal state, and four days after mating they were able to deposit eggs.

Experimental studies of the life-history have shown that with a favourable temperature (about 25° to 30° C.) and suitable moist food such as horse manure, the whole development from the egg to the perfect fly may be accomplished in about nine or ten days and that fourteen days later the second generation may be depositing eggs. In other words, it is possible for the eggs of the second generation of flies to be deposited in slightly more than three weeks after the deposit of the first eggs. With these facts in mind the astonishing fecundity of this insect will be readily appreciated, and the enormous numbers of flies which succeed hot spells of weather will not be a matter of so surprising a character. The relation of the abundance of flies to the presence of breeding places will be referred to in the next chapter.

CHAPTER IV

THE HABITS OF THE HOUSE-FLY

IT would appear to be almost presumptuous to say anything about the habits of the house-fly, and the reader's indulgence is requested.

Of all the habits the most observed and certainly the most important are its feeding habits. We know it as a constant guest at our tables sipping the milk and tasting the sugar and other articles of food. The other articles which it may include in its diet, however, are not generally considered. One does not like to think that the fly now walking round the edge of the cream jug was regaling its impartial palate on the choicest morsels in the dust-bin, ash-pit or garbage can or on more indescribable filth, and yet filth and flies are practically synonymous terms. To that we shall return later.

It has already been stated that the proboscis of the fly is adapted to a sucking function and on this account it cannot take in solid food. How then, one may ask, does it feed on sugar or on dried specks of milk or sputum? In such cases the solid is first dissolved by the saliva of the fly; this saliva is produced in the long salivary glands which open by a single duct into the mouth. As the fluid is produced by the

dissolving of the solid food it runs into and along the
channels of the oral lobes into the mouth and is pumped
up by the powerful pharyngeal pump into the oesophagus through which it runs into the crop. If the insect
is undisturbed while it is feeding, as soon as the crop
is full to the end of its duct the fluid begins to flow
along into the proventriculus and so on to the chyle
stomach. Very frequently, however, the meal is in-
terrupted before the crop is too full and then one of the
objects of the crop may be appreciated. The fly seeks
a conveniently quiet place to digest its meal. The
food passes forward from the crop into the proventriculus which may act not only as a valve closing
the entrance to the chyle stomach, but also as a pump.
From the proventriculus the food passes into the chyle
stomach where digestion begins. The crop undoubtedly serves as a reservoir for the food enabling
the fly to obtain during a short time sufficient food
to last for some time. It calls to mind the function
of the storage stomach of the ruminating animal
which enables such an animal as the cow to obtain a
considerable quantity of food, which it may afterwards
regurgitate and chew at its leisure and when undisturbed. Very frequently the fly regurgitates its food
from the crop in the form of large drops of fluid
which may equal in diameter the depth from front to
back of the head (fig. 8). In some cases these regurgitated drops are deposited as 'vomit' spots on

the surface upon which the fly is resting, such as the
glass pane of the window, when they may be recognised as opaque light-coloured spots as distinguished
from the brown excreta spots commonly known as
'fly specks.' The fly does not always deposit the
regurgitated fluid. In many cases it will regurgitate
a drop of fluid and repeatedly and alternately re-
absorb the drop. One fly which I had under observation was seen alternately and regularly to regurgitate

Fig. 8. House-fly in act of regurgitating liquid food. Such a drop
when deposited forms a 'vomit spot.'

and absorb a drop of fluid eight times, each regurgitation and absorption occupying one and a half minutes.
It is not unlikely that this process may be concerned
primarily with the digestion of the food, as the
regurgitation of the drop of fluid from the mouth
would enable it to become mixed with a further
amount of salivary fluid and further digestion would
be thus facilitated on the absorption of the drop.

Graham-Smith, who has made some interesting observations on the feeding habits of the fly, found that under ordinary conditions the fluid food begins to pass into the ventriculus within ten minutes and in two or three hours, when it has been coloured, it may be found throughout the intestine. One frequently finds that coloured food, such as jam, will remain in the crop several days.

The influence of temperature upon the numerical abundance and the activity of flies is a very close one. The house-fly, although it can withstand subjection to very low temperatures, is extremely sensitive to changes of temperature as the most casual observer must have noticed. A fall of temperature will change flitting activity into stupor. The temperature influences the numerical abundance of flies by affecting the rate of development as will have been noted in the previous chapter. A high temperature therefore is conducive to the production of large numbers of flies by shortening the developmental period; hastening on sexual maturity and stimulating activity and egg laying. With the approach of the cold weather season in October and November the flies seem to disappear in all but the warmest places such as kitchens, restaurants and stables, and even in these places their numbers are decreased.

The question is constantly asked: What becomes of the flies during the winter? Most of them die:

the remainder hibernate. The deaths of many are caused by the parasitic fungus *Empusa* which will be described later, and the rest die of old age. A parallel case is observed in the hive bee. The old worker bees die at the end of the summer, and the young latest born workers live through the winter. The life of a fly in summer is probably not a long one; under the most favourable conditions I have never been able to keep flies alive in captivity for a longer period than seven weeks in summer. Griffith, however, succeeded in keeping a male fly sixteen weeks in captivity, and also obtained four batches of eggs from female flies in captivity. Although female flies under certain conditions of temperature will deposit eggs during the winter months, such a procedure does not appear to be common and the greater proportion of the remnant of flies persisting during the winter months go into complete hibernation. They select some hidden crevice, as for example, behind wood-work or wall-paper for the winter rest and here they remain until the warm days of spring. On exceptionally warm days they will frequently venture out in a somewhat torpid condition, their abdomens being in a collapsed state owing to the fat body having been almost completely absorbed during hibernation. When the fly begins to feed, however, the shrunken alimentary tract soon regains its normal size and with the development of the reproductive

organs the abdomen assumes its usual plump appearance.

The wide distribution of the house-fly is well
known to all who travel, although it should be noted
that the reports of all observers cannot be relied
upon. There are a large number of species of flies
which closely resemble the house-fly in size and
appearance and these are easily mistaken by the un-
trained observer for house-flies, just as the biting
stable-fly is mistaken for an ill-tempered or hungry
house-fly to the injury of the latter's reputation.
Consequently, only an examination of actual specimens
by entomologists can be relied upon to determine
the distribution of the house-fly. As it has followed
civilised man over the entire surface of the globe its
distribution is coincident with man's. Affected as its
reproductive powers are by temperature, it is only
natural that we should find them most numerous in
the warmer regions or in countries which have a high
summer temperature. Another factor which influences
their abundance and distribution is the prevalence
of insanitary conditions; of two places with equal
temperature the locality having the more insanitary
conditions will have the greater number of flies. As
illustrating the abundance of flies in relation to the
breeding places, Major Faichne in India reared about
four thousand flies from one-sixth of a cubic foot of
ground from a latrine and as many as 500 from a

single dropping of human excreta. Herms in California obtained 10,282 larvae of the house-fly from fifteen pounds of samples taken from five different parts of a manure pile after four days exposure. The whole manure pile weighed about 1000 lbs., and on a conservative estimate it would contain over 455,000 maggots. From 1¾ lbs. of manure collected at random 2561 pupae of the house-fly were obtained.

The variability of its distribution locally is of equal interest to its world wide distribution. Locally, its distribution is almost always dependent upon the presence of breeding places as will be shown in discussing preventive measures. This leads us to the consideration of another matter, and that is the question of flight.

One is frequently asked—How far can a fly travel? Not only in the case of the house-fly but in the case of all small winged insects of equal frailty, this question cannot be definitely answered. A number of factors govern the flight potential, as one may term it, of the fly. Included among such factors are, the nature of the locality, whether it is urban or rural; meteorological conditions such as wind and rain; the altitude of the fly, and so forth. This question is of great practical importance in connection with the disease-germ carrying powers and the location of breeding places and on this account a number of interesting experiments have been carried out. In 1906 M. B. Arnold caught and marked 300 flies by

means of a spot of white enamel on the back of the
thorax; these were liberated and five out of the three
hundred were recaptured at distances varying from
30 to 190 yards from the point of liberation. More
recently Copeman, Howlett and Merriman (1911), in
connection with the Local Government Board's inquiry on flies as carriers of infection, conducted some
experiments on the range of flight of flies. The scene
of these experiments was the neighbourhood of a
small village, Postwick in Norfolk, where a plague of
flies occurred. The flies were caught in a net and
were marked by being placed in a paper-bag containing finely powdered coloured chalk of which red
and yellow proved to be the best colours. The
coloured flies were liberated and were recovered from
human habitations in the district at various periods
within forty-eight hours and at distances ranging
from 300 to 1700 yards, that is practically a mile,
from the point of liberation, the location depending,
apparently, upon the direction of the prevailing winds.
The locality in which these experiments were carried
out was of a rural character and consisted of open
country. Flies have been found by me at an altitude
of 80 feet above the ground and this fact has a
significant bearing on the relation of wind to the
range of flight of these insects.

Howard (1911) records an experiment of J. S. Hine
who caught 350 flies and marked them with gold

enamel before liberation. These marked flies were repeatedly observed about dwellings at distances varying from 600 to 1200 yards from the point of liberation up to the third day.

In the summer of 1911 further experiments were carried out under my direction by G. E. Sanders to study the range of flight under city conditions. Flies were reared and marked before liberation by means of an alcoholic solution of rosalic acid with which they were sprayed. The presence of one of these marked flies on a sticky fly-paper is indicated by the fact that when the paper bearing the flies is dipped into water made slightly alkaline the marked fly produces a scarlet colouration. About 14,000 marked flies were liberated on an island in the middle of the Rideau River running through a section of the city of Ottawa. The district contained several very large collections of refuse and horse manure producing millions of flies, consequently the recovery of 172 marked flies was remarkable. They were caught in the kitchens of houses up to a distance of 700 yards in a direct line from the point of liberation, which under city conditions indicates that they had travelled a greater distance.

The bearing which the above data, as to the distance which flies are able to travel in town and country, has upon such practical questions as the placing of stable refuse and other rubbish and the location of sources of infection will be apparent.

CHAPTER V

OTHER SPECIES OF FLIES FOUND IN HOUSES

FREQUENT mention has already been made of the occurrence of other species of flies such as the Lesser house-fly (*Fannia canicularis*) and the Stable-fly (*Stomoxys calcitrans*) in houses, indicating that, while the house-fly, as both its popular and scientific names imply, is the truly 'domestic' fly, there are other species which from time to time occur in houses. Some of these such as the Lesser house-fly are co-inhabitants of houses, others such as the Stable-fly and 'Blue-bottle' or 'Blow-fly' can only be regarded as visitants as they normally lead an open-air life and visit houses only to seek shelter or food.

If a census is taken of the flies occurring in human habitations it is found that the house-fly constitutes by far the greater proportion. In a collection of flies made in different cities in the United States, in rooms where food supplies were exposed, Howard (1900) found that out of 28,087 flies, 22,808 or 98·8 per cent. of the whole number were *Musca domestica* and of the remainder the Lesser House-fly (*Fannia canicularis*) constituted the greatest proportion. In 1907, I found that out of a collection of 3856 flies caught in a variety of places such as restaurants, kitchens, stables, bedrooms, etc., in Manchester, 87·5 per cent. were *Musca domestica*,

11·5 per cent. were *Fannia canicularis* and the rest were such other species as the Stable-fly, the Blow-fly, etc. Hamer found in an investigation in London that in a collection of 35,000 flies caught on four fly-papers exposed in similar positions, in addition to the house-fly, 17 per cent. were *Fannia canicularis*, less than one per cent. were blow-flies and less than one per cent. were another species *Muscina stabulans*. Out of 8553 flies caught in six different localities in Manchester, Niven found 8196 were *Musca domestica*, 293 were *Fannia canicularis* and 64 were other species. In a collection of 24,562 flies made in the city of Birmingham Robertson found the different species of flies in the following proportions : *Musca domestica*, 22,360, 91 per cent. ; *Fannia canicularis*, 1154, 4·7 per cent. ; the blow-fly (*Calliphora erythrocephala*), 840, 3·4 per cent. ; and other species, 218 or ·9 per cent. As a general rule *Musca domestica* constitutes more than 90 per cent. of the total fly population of a house. The proportions of the other species vary, the variation depending largely on the situation of the house whether it is urban, suburban or rural.

In India two species of flies are found which are closely allied to *Musca domestica* and on account of the general similarity of their breeding habits and close resemblance they are frequently mistaken for the true house-fly. They are *Musca domestica* sub sp. *determinata* and *Musca entaeniata*.

THE LESSER HOUSE-FLY, *Fannia canicularis* L.
(Fig. 9.)

From the foregoing figures it will have been noted that this species occupies the first place of importance after *Musca domestica* as a frequenter of human habitations. Mention has already been made of the fact that it is not infrequently mistaken for a 'young'

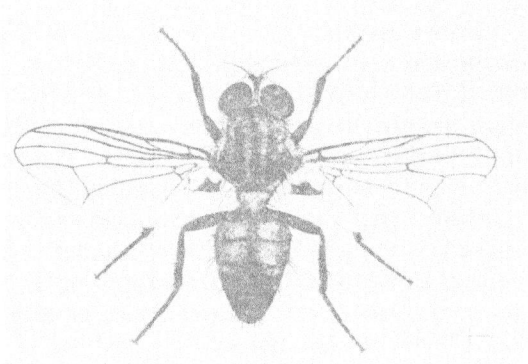

Fig. 9.   The Lesser house-fly, *Fannia canicularis*.   Male.

house-fly by those unacquainted with the nature of the growth of these insects. It belongs, however, to a distinct family of flies which are characterised by the fact that the fourth longitudinal vein of the wing goes straight to the margin of the wing and does not bend upward at an angle as in the house-fly.

This species is more slenderly built than *Musca domestica*. The back of the thorax is striped, rather indistinctly in the male but more plainly in the female, with three dark lines. The abdomen of the male is narrow and tapering compared with that of *Musca domestica* and the three or four basal segments have each a pair of golden translucent areas at the sides which are readily seen by the transmitted light when the fly is on a window pane.

This species appears in the house before the true house-fly and may be found usually in May and June. Later it is displaced by *Musca domestica*. The females are often captured out of doors, while the males accompanied by a varying number of females may frequently be observed flying around chandeliers, etc., in the living rooms and bedrooms of houses in a characteristic jerky and hovering manner. An old writer has referred to these flies as uniting in 'little social parties that circle for hours in a sober uniformity of flight below the ceilings of our chambers.'

The breeding habits of *Fannia canicularis* are, on the whole, similar to those of the house-fly. The larvae breed in decaying and fermenting vegetable and animal matter and also in the excrement of various animals including man. The eggs are white and similar in shape to those of the house-fly. The larvae (fig. 10) however are totally unlike house-fly larva. They are compressed dorso-ventrally. The surface of

the body is rough and spinous.   There is a double
row of spiny processes along each side of the body to
which the dirt adheres giving the larvae a very dirty
appearance which is enhanced by the dirt covering
the body and other small processes.   The full-grown
larva measures almost a quarter of an inch in length.
The larval life may extend from one to several weeks
and two to three weeks are spent in the pupal state.

Fig. 10.   Larva of Lesser house-fly, *Fannia canicularis.*
a.sp., anterior spiracles; *p.sp.*, posterior spiracles.

THE LATRINE-FLY, *Fannia scalaris* F.

Investigations during recent years would appear
to indicate that this species should be considered
with the house-fly.   Although it has rarely been found

by me in houses nevertheless its breeding habits indicate that its relationship to man may be of some importance from time to time. Owing to its general similarity to the Lesser house-fly it is often confused with this species. It differs from it, however, in several respects. The knee joints (tibiae) of the middle pairs of legs of *Fannia scalaris* are each provided with a distinct tubercle, and the abdomen is black overspread with bluish grey. Both the adult fly and the larvae of the Latrine-fly are slightly larger than those of the Lesser house-fly.

The larvae breed more generally in excrementous matter and are very commonly found in large numbers in human excrement, both in the pure state and when mixed with earth or ashes. They will also feed on decaying vegetable substances. In a series of experiments the larvae emerged from the eggs in eighteen hours after they were deposited and completed their growth in six to twelve days; the pupal stage lasted about nine days.

While the larva of the Latrine-fly has a general resemblance to that of the Lesser house-fly, a close examination of specimens which have been deprived of their dirty covering will show marked differences. In shape the larvae are similar but the lateral appendages of the larvae of *Fannia scalaris* are feather-like in appearance forming a fringe around the body of the larva. The feather-like nature of these lateral

appendages which were formerly mistaken for gills
such as we find in certain aquatic insect larvae, are
no doubt an advantage to the larva living, as it
usually does, in liquid filth.   Its relation to man will
be described in a later chapter.

THE STABLE-FLY, *Stomoxys calcitrans.*
(Fig. 11.)

This species is an out-door fly which sometimes
enters and may remain inside houses.   It loves the
sun and may be found on sunny days alternately
resting or hovering over doors, gates and fences
which are exposed to the full glare of the sun.
Wherever cattle and horses occur this species will be
found in abundance and consequently it is a common
frequenter of farm-yards.

Reference has already been made to its similarity
to the house-fly, which is responsible for a biting habit
having been attributed to the latter insect which is
unable to bite.   A close examination of the Stable-fly,
especially with the aid of a small lens, will disclose
certain well-marked differences, for, although this
insect is nearly related to the house-fly, the biting
habit requires a differently constructed proboscis.
When observed from above this awl-like proboscis,
adapted for piercing and sucking, can be usually seen
projecting forwards horizontally from beneath the

head (fig. 12). A similar proboscis is found in the
Tse-tse flies which are close relatives of our Stable-fly,
but more famous on account of the deadly nature
of the disease-causing organisms they transmit, the

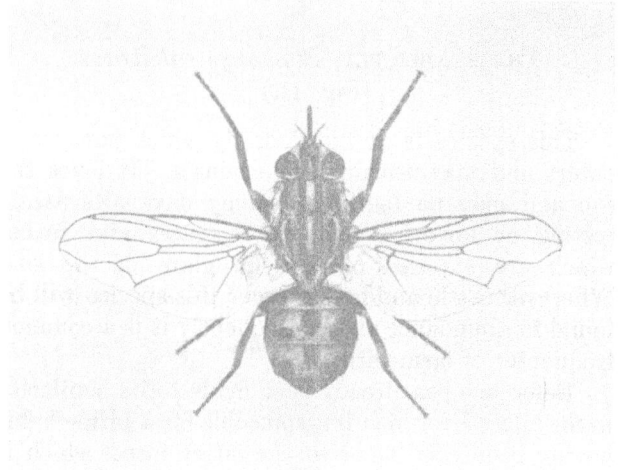

Fig. 11. The Stable fly, *Stomoxys calcitrans*. Female. Note the
awl-like piercing proboscis projecting forwards from beneath the
head.

microscopic Trypanosomes. The body of the Stable-
fly is a little larger and more robust than that of the
house-fly. The colour is brownish with a greenish
tinge and on the dorsal side of the thorax are four

dark longitudinal stripes. The golden tinge of the
anterior end of the median light-coloured stripe is
very characteristic. The abdomen is broadly robust
and bears a number of brown spots.

In the country I have commonly found it from
July to October, but specimens have also been found

Fig. 12. Head of the Stable-fly, *Stomoxys calcitrans*.

on the windows of a country house as early as March
and April, and Howard states that John B. Smith
found them abundant in the house at the end of
October. It is normally an out-door insect and its
reasons for entering houses, except occasionally to
seek shelter in case of rain, as I have observed, is

difficult to ascertain. Being a blood-sucking insect it is not attracted by the same means as the house-fly, but when it has entered a house it usually takes advantage of its opportunites for a variation in diet, and the house-fly is blamed.

It breeds readily in decaying vegetable refuse, especially if it is warm on account of fermentation. It has also been reared from the excrement of various domestic animals such as the horse, cow and sheep. From fifty to seventy eggs are laid, the eggs being white and of the same size as those of the house-fly. The larvae of *Stomoxys calcitrans* and of the house-fly are similar in general appearance, but the larva of the former has a more shiny and translucent appearance and the posterior spiracles or breathing pores are different. Newstead found that the whole larval stage lasted from fourteen to twenty-one days and the pupal stage from nine to thirteen days. The whole life-history may be complete in twenty-five to thirty-seven days. Some individuals may pass the winter in the pupal state.

Owing to the fact that the feeding habits of *Stomoxys calcitrans* differ from those of *Musca domestica*, in that it does not frequent to so great an extent substances likely to contain intestinal bacilli of a disease-causing nature, it is probably not a serious factor in the carriage of the diseases for the dissemination of which the house-fly is noted. Its blood-sucking habits, however, may be responsible for the occasional transfer of the germs of anthrax from cattle to man giving rise thereby to malignant pustule or other bacillary disorders.

THE BLOW-FLY OR BLUE-BOTTLE,
Calliphora erythrocephala.

No account of the fly visitors of our houses would be complete if this species were omitted. Not by reason of its abundance is it usually known in the house, but rather by its persistent buzz and elusive nature. Its large size and hairy dark-blue body makes it easily distinguishable.

It usually enters houses in search of material upon which to lay its eggs as the housewife who inadvertently leaves the meat exposed or uncovered can testify when it is subsequently discovered in a 'fly-blown' condition or worse. The Blow-fly larvae or maggots, which are sometimes called 'gentles,' feed on fresh or decaying meat. Sometimes they feed on the living animal, as for example on the back of a sheep or on a trapped animal. I have found a living mass of small larvae on the broken leg of a living rabbit which had not been caught twenty-four hours. This insect is remarkable for the large number of eggs which the female fly deposits. Portchinski has found as many as 450—600 eggs deposited by a single fly. In a series of investigations, using fresh meat (rabbit) the shortest time of development which I found was as follows : The eggs hatched from eight to twenty hours after deposition ; the larval life composed of three stages, was passed in seven and a half to eight days and the pupal stage lasted fourteen days. The development was thus complete in slightly more than three weeks but I believe that under certain conditions this time might be shortened. The full-grown larvae may be distinguished from the larvae of the house-fly, not only by their large size but more especially by the fact that the posterior end is surrounded by twelve tubercles. The anterior spiracles have nine lobes.

These flies frequent fresh human faeces, as may be observed in insanitary places, and also fruit and meat which is often openly exposed for sale. These facts lend some support to the possibility of the Blow-fly occasionally serving as a disseminator of intestinal bacilli ; their flesh-seeking habits also render them liable, as experiments have shown, to transmit anthrax bacilli if they have access to infected fresh flesh.

THE CLUSTER-FLY, *Pollenia rudis* Fabr.

Not infrequently, especially in country houses or in houses surrounded by ivy or other creepers, swarms of rather large sluggish flies will appear in the spring and sometimes in the autumn, and these flies are usually mistaken by the casual observer for house-flies. I have found them entering the sunlit window of a bedroom in a country house in large numbers and swarming over the window panes. An examination of these flies will indicate well-defined differences from the house-fly. They are slightly larger in size and darker in colour with a yellowish tinge; when at rest their wings are folded more closely together over the back than is the case with the house-fly. Their slow movements are also a distinguishing characteristic.

The breeding habits of the Cluster-fly are not so well known as certain of the species already described. Howard records the species as having been reared from excrementous matter and it is also believed to breed in decaying vegetable matter. Keilin describes it as being parasitic on a species of earthworm *Allobophora chlorotica* and states that it pupates in the earth, the pupal stage lasting from 35 to 42 days. The adult flies may be found clustered together in large numbers in out of the way corners and crevices, especially in rooms seldom occupied.

Muscina stabulans (Fall).

No popular name has been used in the present instance to designate this insect. The reason for this omission is simple, namely, that I am unable to find or to give it one. It has been called the stable-fly on account of the specific name *stabulans* but as it does not frequent stables to any marked degree this name is not strictly applicable, apart from the fact that it is more usually and more correctly used for *Stomoxys calcitrans*. In naming animals and plants in the majority of cases the scientific names are descriptive, describing some structural feature or distinctive habit, and the popular name is usually of the same character. This is well-illustrated in the case of *Musca domestica*. The popular name may be a free or literal translation of the scientific (specific) name. On the other hand the scientific name may describe a structural character and the popular name another character or habit as in *Caliphora erythrocephala*, the Blow-fly or Blue-bottle; the second part of the scientific name referring to the red colouration of the insect's cheeks or genae. Now in the case of a person a name does not usually describe him in the same sense (we cannot discuss here the mental descriptive picture which a person's name may conjure up in our minds and its further bearing on his family relationships), it is more of a distinction. Why then should we not in cases where it is impossible to find a descriptive name of any degree of accuracy adopt a distinguishing name, especially when the scientific name provides us with one? It is as simple, and certainly more accurate, to call this insect popularly *Muscina stabulans* as to manufacture an unwieldy and inaccurate popular name.

To return to our subject. This fly is larger than *Musca domestica* but so remarkably similar to it in general colouration that it is easily mistaken for a large house-fly. I have usually found it in houses together with the Lesser house-fly in the early summer before the house-fly appears in large numbers. It breeds on all kinds of decaying vegetable substances and also feeds on growing vegetables. The larvae have also been found in excrement and in the remains of rotting insects. The habits of this insect, which resemble to some extent those of the house-fly, make it of some economic importance. Further, its larvae have been found to occur in the human intestine.

There are a number of other species of flies found more or less frequently as visitants in human habitations and for an account of these the reader is referred to the more comprehensive works on this subject.

CHAPTER VI

THE PARASITES AND NATURAL ENEMIES
OF THE HOUSE-FLY

LIKE other living creatures the house-fly has its enemies and parasites. Small though it may seem yet it is subject to the attack of no inconsiderable number of parasites ; some of them are other insects, others are mites, minute worms, and unicellular or protozoal organisms. One of the most important parasites from a practical point of view, that is, considered as a factor in the natural control of this insect, is the parasitic fungus *Empusa muscae*.

THE HOUSE-FLY FUNGUS, *Empusa muscae*.

In the early autumn flies may frequently be found attached, in so life-like a position that they are generally believed to be living until they are touched, to the window pane, to the wall or to the ceiling. When such flies are touched, however, one may be surprised to find that they do not move but are dead. Gilbert White says: 'But as they grow more torpid one cannot help observing that they move with difficulty, and are scarce able to lift their legs which seem as if glued to the glass ; and by degrees many do actually stick on till they die in the place.' If one of these flies be examined it will be observed that the body is swollen and white bands occur between the expanded segments of the abdomen. Such a fly has fallen a victim to this fatal disease caused by the fungus *Empusa muscae*. The whole of the internal organs of the fly are destroyed and had it been allowed to remain in its fixed position a little longer a ring or halo of white dust-like spores would have been noticed surrounding the victim, in fact, flies killed by this parasite are usually first detected by the ordinary observer owing to the presence of the white ring of what are really fungal spores or seeds (see fig. 13).

The appearance of this disease is not usually noticed until the beginning of July and from that time it may be more commonly observed until October and November when the numbers of flies decrease. The mortality caused by the *Empusa* has frequently a marked effect on the reduction in numbers of flies in the autumn. This species of fungus belongs to a large class called the Entomophthoreae, which, as their collective name would imply, are confined to the insects. There are a number of species of *Empusa* one of which *Empusa gryllii* is sometimes an important factor in the control of grasshoppers in North America. *E. aulicae* occasionally causes a high rate of mortality among caterpillars.

The life-history of the *Empusa* is of very great

interest and although many attempts have been
made no one has yet succeeded in working out the

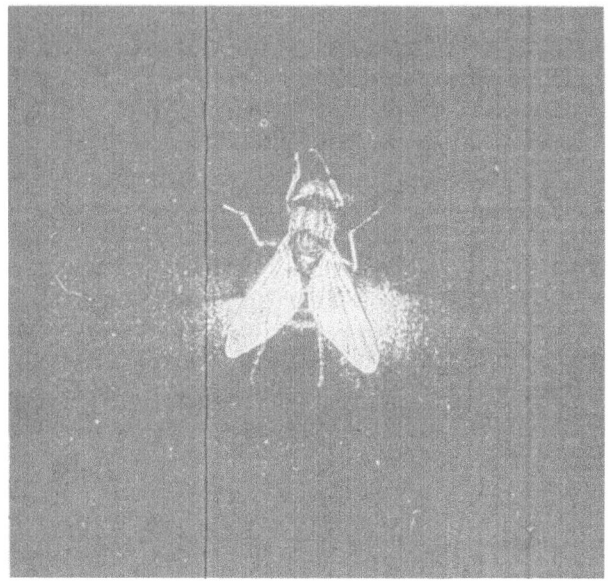

Fig. 13. House-fly attacked by fungal disease, *Empusa*, showing
surrounding spores in the form of white dust. (*Photo by
H. T. Güssow.*)

whole life-cycle. When the fly is found attached to
the window pane or chandelier in a life-like though
lifeless condition the disease has developed very

considerably, and we must go further back to an occasion several weeks earlier when one or more spores or germs of the *Empusa* settled by some means or other on the body of the insect and, held in place by the abundant hairs, remained there. From this spore a small filament or hypha grows out and pierces the chitin probably by means of a dissolving action. Having pierced the body wall it enters the fat body and gives rise to a large number of small spherical bodies. These divide as they are formed and being free are carried along in the blood to all parts of the body. These little bodies in turn produce fungal threads or hyphae which ramify all through the body, gradually destroying all the muscles and internal organs of the fly until it is almost a solid mass of interwoven fungal hyphae (fig. 14). When this stage is reached the hyphae pierce the softer parts of the body wall, namely, the membranes between the segments of the body, and produce short, thick spore-bearing stems called conidiophores which are closely packed together in pallisade fashion forming the white ring observed externally between the segments. Each of these conidiophores produces a bell-shaped spore or conidium at its apex. These spores are very minute, measuring about one-thousandth part of an inch long. When they are ripe they are discharged and may be thrown some distance from the fly. Mr H. T. Güssow has found that spores

may be discharged to a distance of 2⅘th inches from
the fly. Millions of these white spores or conidia
shot off in this manner from the white ring may be
seen round the dead fly. The object of this method
of spore discharge is to enable the spores to reach
another fly. If the spore after being discharged by
the conidiophore does not reach another fly it

Fig. 14. Section of abdomen of house-fly attacked by *Empusa*
    showing the whole interior destroyed by fungal threads *f*. *C.*
    conidiophores or spore-producing threads.

produces a small conidiophore which in time gives
rise to another spore, and if this spore is as
unfortunate as its parent another may be produced.
The repetition of this process is to enable the spores
to travel as far as possible from the original point of
production with a view to settling upon the body of
a hitherto healthy fly. In the event of this happening

the spore develops in the manner already described. From this it will be gathered that such a method of reproduction will result in a considerable proportion of flies being infected where they are abundant.

The question will arise in the reader's mind, no doubt, as to how the *Empusa* is carried on from one year to the next in view of the fact that under ordinary conditions the flies practically disappear during the winter months. It is here that the gap in our knowledge of the life-cycle of this fungus is encountered. It has been suggested that resting spores or spores which remain in a quiescent state are produced. Carrying this idea still further, it is supposed by some that the larvae become infested with these resting spores, which would naturally occur in places where the dead bodies of diseased flies were found. From what we now know as to the ability of insect larvae to ingest micro-organisms and transmit them to the adult mature insects there is a certain degree of probability in this idea, but we have no experimental evidence as yet in support of it. We do know, however, that the disease reappears in the following summer.

CHELIFERS OR 'FALSE SCORPIONS.'

Not infrequently flies are found having small lobster or scorpion-like creatures attached, generally to their legs but occasionally to their bodies. These small animals, measuring about one-tenth of an inch in length, are related to the spiders and belong to an order *Pseudo-scorpionidea.* They are also frequently called Chelifers on account of the pair of chelate or pincer-like appendages which they possess and which

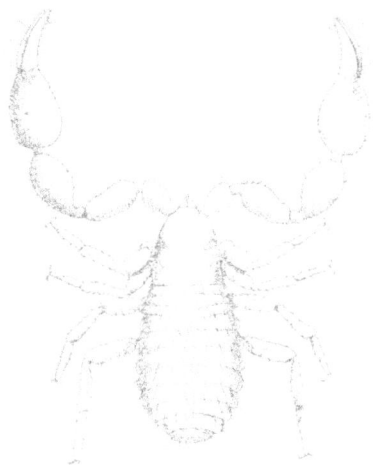

Fig. 15.   Chelifer, *Chernes nodosus*, which infests house-flies.

enable them to cling firmly to the fly. They are yellowish-brown in colour. The head and thorax are united in a single segment and the abdomen bears a number of hairs. In addition to the large and powerful chelae they possess, like all the members of the

spider group, four pairs of legs. Their habits vary
somewhat according to the species; *Chernes nodosus*
(fig. 15), which is the species usually found attached
to the fly, commonly lives among refuse such as
decaying vegetation, manure heaps and hot-beds.
In view of the habits of the house-fly, the frequent
occurrences of this chelifer upon its legs will be readily
understood, since by means of its chelae it is enabled
to attach itself to any unwary fly which may be either
emerging or wholly emerged from a pupa bred in the
rubbish heap or may be visiting the heap for the
purpose of depositing its eggs.

There is considerable difference of opinion as to
the chelifer's object in clinging to the fly. On the
one hand it is held that the object is one of dispersion
and that the chelifer merely clings to the fly in order
to be transported to a new situation. On the other
hand certain observers maintain that the chelifer is
predaceous or parasitic. The chelifers feed upon
small insects and mistaking the size of the fly they
may seize it and be carried away; should the fly die
they might feed upon it, and when a number of
chelifers are attached to a single fly, as is sometimes
the case, the fly's life will no doubt be shortened.
There are one or two records of the chelifers fixing
themselves on to the fly in a parasitic manner.
Whether they are parasitic must remain uncertain
pending the adduction of more evidence, but it is

obvious that the relationship of the chelifer to the fly will result in the former's dispersal.

In addition to chelifers, house-flies may bear small reddish or brown mites or *Acari*. As early as 1735 De Geer observed small reddish mites on the head and neck of house-flies. These reddish mites are often the six-legged larvae of a species of mite belonging to a genus *Trombidium* which, by means of their sucking mouth parts, suck the juices of the fly.

Flies emerging from pupae in a rubbish heap or hot-bed will frequently be found to be carrying numerous small brownish mites. Many of these belong to a group *Gamasidae* which are rather flat and broad mites and their larvae occur in large numbers in such situations as rubbish heaps, etc. To these immature forms the fly serves as a most convenient transporting agent and assists in the emigration of the mites to new fields. Some of these forms are parasitic as I have found them firmly attached by their mouth parts to the under-sides of flies (fig. 16).

In some cases flies serve as the means of transporting the mites, which are destructive to cheese and other foods, in an unusual and interesting manner. Under ordinary favourable conditions, when the food supply is abundant, these mites reproduce with enormous rapidity, the young mites developing in a very short time into adults. When the food supply becomes scarce or other unfavourable conditions prevail, instead of passing through the usual stages of development, the almost fully-grown mites develop hard protective cases or shells into which they can

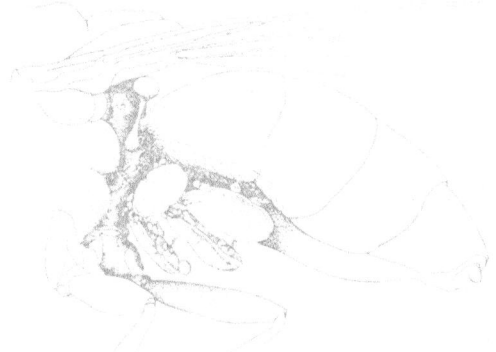

Fig. 16.   Mites attached to abdomen of Lesser house-fly.

draw themselves for protection.   This stage is known as the hypopus and is in reality a migratorial stage. These hypopi, as they are called, attach themselves to flies and are carried away from the unfavourable conditions, under which probably most of the older mites together with the youngest have perished, to other places where they may encounter food.   In the light of this mite-carrying power of the house-fly, the possibilities of food such as cheese, ham, etc., becoming infested with mites will be appreciated perhaps to a greater extent than hitherto.

THREAD-WORM OR NEMATODE PARASITE OF THE HOUSE-FLY

If a piece of damp soil, decaying humus or rotting horse manure is carefully examined with a lens one frequently discerns minute white worms wriggling about. These are thread-worms or Nematodes. While some kinds of thread-worms are free living, others are parasitic on vertebrate or invertebrate animals. Certain forms of parasitic thread-worms cause serious diseases such as Filariasis or Elephantiasis, Ankylostomiasis, etc. In 1861 Carter discovered minute nematodes in the house-fly. They were found chiefly in the proboscis and he stated that every third fly contained twenty or more parasites, to which he gave the name *Filaria muscae*. Other observers including myself have found these thread-worms in the house-fly and the species is now known as *Habronema muscae*. In Italy it was found that at certain seasons as many as 20 to 30 per cent. of the flies were infested with this parasite. It is a thin whitish worm tapering off at both ends and measuring less than one-tenth of an inch in length, and is the larval stage. It is most frequently found in the head and proboscis though it may occasionally occur in the thorax or abdomen. Howard records some observations of Ransom who found nine out of thirty-four flies, most of which were caught at Washington, D.C., infested with the parasites; in seven out of nine flies the parasites occurred in the head and proboscis and the number of parasites varied from one to seven (in one case there were six in the head and proboscis and one in the thorax). The possibility of house-flies carrying pathogenic nematodes in tropical countries should not be lost sight of, as it would be an easy matter for such nematodes if contained in the head and proboscis to be transferred into the food. Ransom (1911) has since found the complete life-cycle of this parasite as a result of the examination of the stomachs of horses. The embryo nematodes are excreted with the horses' faeces and enter the bodies of the fly larvae. When the flies emerge from the pupal stage the nematode larvae are full-grown and their further fate waits upon the swallowing of the infested flies by a horse, in which event the life-cycle of the parasite is completed by the growth of the worms to maturity in the alimentary canal of the horse.

PROTOZOAL PARASITES.

As long ago as 1878 a protozoal parasite of the house-fly was figured by Stein. This microscopic unicellular parasite, now known as *Herpetomonas muscae-domesticae*, was studied in 1904 by Prowazek who gave a detailed account of its development. During the last fifteen years our knowledge of these parasitic flagellates has been increased enormously and they have been found widely distributed throughout the animal kingdom. This quest of the parasitic flagellate has been largely due to the fact that several species belonging to the group *Trypanosoma* have been found to be causative organisms of certain fatal diseases to man and domestic animals. Insects have been found to harbour a number of species. The Tse-tse flies carry the Trypanosomes of Sleeping Sickness, Surra, Dourine, etc. ; the bed-bug has been recently shown by Patton to transmit the causative organisms of the serious tropical disease Kala-azar or black fever. The occurrence of a species of *Herpetomonas* in the house-fly is of interest despite the fact that the non-biting habits of the fly will probably preclude its ever being discovered to be a carrier of a flagellate blood parasite. Patton has made in 1908 and subsequently a careful study of this species together with other species and his

observations have been confirmed. The full-grown
flagellate measures up to one five-hundredth of an
inch in length; its body is lancet-shaped and contains
two nuclear structures. A stout whip-like flagellum
arises from the anterior end of the organism. This
flagellum was described by Prowazek as being double
but Patton and subsequent investigators have shown
that the double state is a phase in the division of
the parasite. There are three stages in the developmental history; the pre-flagellate, flagellate and
post-flagellate. The pre-flagellate stage occurs in the
mid-gut of the fly; in this stage one finds minute round
or slightly oval bodies which divide by simple longitudinal division or multiple segmentation to form a
large number of individuals. These soon develop into
the flagellate stage by the formation of a flagellum
and the elongation of the body of the organism.
In the flagellate stage the organisms divide longi-
tudinally. In starved flies the flagellate forms collect
in the rectal region and attach themselves to the gut
epithelium in rows. Later, some of them begin to
shorten, divide and become rounded, and after losing
their filaments they form cyst-like bodies which
pass out with the faeces, and falling on to surfaces
frequented by flies serve to infect other flies by
being taken up in their proboscides.

Patton found that one hundred per cent. of the
flies were infected with this parasite in Madras. At

Rovigno, Prowazek found about eight per cent. of
the flies infected.

A species of Crithidia, which is another group of
flagellates, has been described by Werner from the
house-fly, but Patton and Strickland are probably
correct in their suggestion that this is a stage in the
life-history of *Herpetomonas muscae-domesticae*.

INSECT ENEMIES.

In common with most species of insects the
house-fly has its insect enemies. Some of these are
parasitic while others are predaceous and, like birds,
prey upon them.

The parasitic insects affecting the house-fly belong
to two large families of the Hymenoptera which order
of insects contains the greatest number of parasitic
insects. These families the *Cynipidae* and *Chalcidae*
are mostly very small four-winged insects. Most of
the insects belonging to the Cynipidae are gall-
makers and produce the well-known galls upon
plants. The sub-family *Figitinae*, however, are
parasitic on insects, chiefly on flies, and one or two
species of *Figites* have been recorded as parasitic of
the house-fly.

The other family, the *Chalcidae*, contains the
larger number of parasites of the house-fly and

although it is only within the last few years that
anything approaching a thorough study of the chalcid
parasites of the house-fly has been made, nevertheless,
owing to the investigations of A. A. Girault and
G. E. Sanders in Illinois, U.S.A., we have considerable
information concerning a number of these parasites.
Of the several species of parasites which they have
studied reference will be made to one only, namely
Nasonia brevicornis. This minute parasite is of
a metallic dark brassy-green colour with garnet
coloured eyes. It is very sluggish in its movements.
The female measures 1 mm. to 2·30 mm. in length, and
the male is about one-third smaller in size. These
female chalcids deposit their eggs in the pupae of
the house-fly. The average length of the life-cycle
of the parasite is a little over three weeks and it was
found that the parasite is able to hibernate as a full-
grown larva in the puparium of the host, changing to
a pupa early in the spring and emerging shortly
afterwards. Another genus *Spalangia* is a parasite
of the house-fly both in Europe and in North
America.

Of the predaceous enemies of the house-fly the
common wasp is not the least. The robber-flies or
Asilidae, a family of rather long-bodied hairy two-
winged flies, may be seen frequently seizing and
eating house-flies. In the larval stage they are
frequently devoured by the larvae and adults of

several families of beetles such as the ground beetles (*Carabidae*) and rove-beetles (*Staphylinidae*). Compared with the parasitic enemies the control which the predaceous enemies of the house-fly effect is extremely slight.

PART II

THE RELATION OF HOUSE-FLIES TO DISEASE

CHAPTER VII

THE CARRIAGE AND DISTRIBUTION OF MICRO-ORGANISMS BY FLIES

' The kingdom is much pestered with flies in summer; and the odious insects, each of them as big as a Dunstable lark, hardly gave me any rest while I sat at dinner with their continual humming and buzzing about mine ears. They would sometimes alight upon my victuals and leave their loathsome excrement or spawn behind, which to me was very visible.'

SWIFT.

THE verification of the belief that the house-fly is able to carry and disseminate micro-organisms including those of certain infectious diseases is one of the notable advances in preventive medicine during the last fifteen years. It is a mistake, however, to imagine, as some appear to do, that the idea of a relationship between house-flies and disease is of recent origin. It is a belief which had its origin hundreds of years ago. The instructions of Moses (Deut., ch. xxiii. 12, 13) would indicate as great an appreciation of the part which flies play in the dissemination of disease as that displayed by modern surgeons-general. From the sixteenth century onwards medical literature contains evidences of this idea and several pages could be devoted to references to these suggestions, but only a few of the later references will be considered. In 1869 Raimbert experimentally proved that the house-fly and blow-fly were able to transmit the anthrax bacillus. Lord Avebury, writing in 1871, referred to the habits of flies alighting on decomposing substances and carrying impurities especially the secretions of unhealthy wounds. He said that rather than regard them as dipterous angels dancing attendance on Hygeia we should look upon them as winged sponges spreading hither and thither to carry out the behests of foul contagion. Leidy in the same year (1871) declared his belief that flies were responsible for the spread of hospital gangrene and wound infection during the American Civil War. In 1880 Laveran showed that flies would carry on their proboscides and legs the infectious discharge of conjunctivitis or ophthalmia. Further reference will be made to other early work in the succeeding chapters ; sufficient has been said, however, to indicate that this belief in the disease carrying powers of the house-fly is not one of recent origin. Nevertheless, there is no doubt that the attention which this subject deserved was not devoted to it until after the discovery, which was made as a result of the investigations into the United States military camp conditions during the Spanish American war in 1898 and the British camp conditions in the Boer War a year or two later, that flies under the necessary conditions were the most important factors in the dissemination of typhoid fever.

There are several ways in which those insects which play a part in the dissemination of disease serve as intermediaries. In the case of the mosquito and malaria, the malarial organism, which is a blood parasite, undergoes a definite series of developmental changes in the body of the mosquito before the resulting malarial spores are injected into another human being by the piercing and sucking proboscis of the mosquito. In the case of transmission of the Trypanosome blood parasites which are carried by the Tse-tse fly from one mammal to another, some of these parasites are mechanically transferred on the proboscis of the Tse-tse fly, which resembles the proboscis of our own Stable-fly, in the process of piercing the skin and sucking the blood. Some of the parasites, however, undergo certain developmental changes in the gut of the fly, and as in the case of the mosquito, may enter the salivary glands. In the case of a non-blood-sucking insect, such as the house-fly, the method of transmission is different.

Micro-organisms carried by the house-fly cannot reach the circulatory system in the same way owing to

the inability of the fly to pierce the skin with its
proboscis. To penetrate the system the organisms
have to reach it in most instances by way of the
mouth, that is, on the food. There are, of course,
other micro-organisms, such as the causative organisms
of ophthalmia or conjunctivitis, which do not need to
pass through the system but whose evil effects are
localised externally. In all these cases the method of
transferring the micro-organisms is mechanical and
direct, by which is meant that no developmental

Fig. 17. Foot-joints of leg of house-fly to show bristly character.

change of the micro-organisms takes place in the fly
during the transference of the organisms.

Micro-organisms are transferred either externally
or internally by the fly from the source of infection.
As a means of transference the body of the fly is most
excellently adapted being thickly clothed with hairs
or setae of varying degrees of length. Its legs, which
chiefly come into contact with the infective materials
upon which it walks, resemble miniature brushes
(see fig. 17), from which no cleaning can remove the
organisms once these appendages have been defiled,
with the result that they contaminate whatever

substances they subsequently visit within a certain length of time. The mechanical transference by the appendages and body of the fly, however, may not play so important a *rôle* as the transference due to the feeding habits of the fly. In feeding on infected matter, as flies invariably are accustomed to do, whether it is excreta infected with typhoid bacilli, tubercular sputum, or a purulent discharge, the micro-organisms are taken into the gut of the fly where they are able to remain a greater length of time than if they were on the body or appendages, the danger of dissemination being removed to a very great extent. From the gut of the fly, the organisms may reach our food in either of two ways, namely, by de-faecation and by regurgitation or vomit spots. In both cases the faecal spots or vomit may contain a certain number of the micro-organisms. An important series of experiments by Graham-Smith demonstrated that flies, artificially fed and kept in captivity, were able to contaminate milk upon which they fed, for several days. Total immersion of the fly also caused infection.

All this leads to the one conclusion that nothing can prevent a fly which has had access to matter containing micro-organisms from carrying those micro-organisms and infecting fresh matter which, in our case, is usually food. Further, one is able to make the assertion that, owing to the possibilities of becoming infected and the fact that the fly is continually coming in contact with micro-organisms in the materials it frequents for the purposes of food or breeding, no fly is free from micro-organisms or germs from the time it draws itself out of the puparium to enter the winged state until its death.

The fact that all house-flies, wherever they may be caught, normally carry a varied collection of fungal and bacterial organisms may be practically demonstrated by any one having a knowledge of bacteriological methods. The accompanying figure (fig. 18) illustrates an agar slope culture bearing numerous colonies of bacteria and moulds obtained by allowing a fly caught in my laboratory in January (1910) to make a single journey over the culture slope.

Some experiments carried out by Mr H. T. Güssow in the summer of 1908 are of great interest illustrating the fact that flies are not only normally infected with micro-organisms of various kinds but that the number and variety of micro-organisms vary and are affected by the places frequented by the flies. Three flies, *A*, *B* and *C*, were caught at random ; *A* was caught in the living room of a house (Norwood, London) ; *B* was caught out of doors and *C* was caught in the household dust-bin and refuse can. In each case the fly was allowed to walk over a medium of nutrient agar-agar which was afterwards incubated, and at the end of the fourth day the following colonies of fungi or moulds and bacteria were observed, isolated and identified.

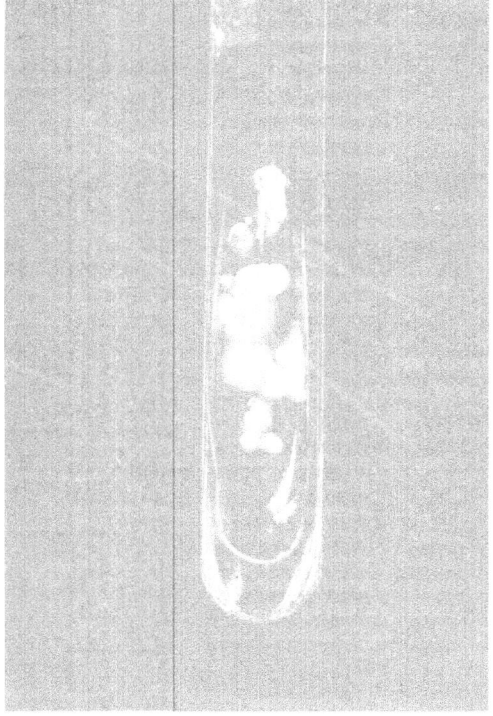

Fig. 18. Agar-agar slope culture of bacteria deposited by a house-fly in a single journey over the culture medium.

A. INDOOR FLY.

Gave 25 colonies of bacteria and 6 colonies of fungi.

Bacteria :

    2 colonies of *Micrococcus ureae*.
    2   ,,   *Bacillus subtilis*.
   11   ,,   *Bacillus coli commune* (Intestinal bacillus).
    2   ,,   *Sarcina lutea* (Intestinal bacillus).
    3 colonies stained by Gram.
    5   ,,   not stained by Gram.

Fungi :

    2 colonies of *Saccharomyces* sp. (Yeast fungus).
    2   ,,   *Penicillium glaucum* (Blue mould).
    1 colony   *Aspergillus niger* (Black mould).
    1   ,,   *Cladosporium herbarum* (Olive green mould).

B. OUTDOOR FLY.

Gave 46 colonies of bacteria and 7 colonies of fungi.

Bacteria :

   18 colonies of *Bacillus tumescens*.
    9   ,,   *Micrococcus pyogenes aureus*.
    2   ,,   *Sarcina lutea* (Intestinal bacillus).
    1 colony   *Sarcina ventriculi*.
    4 colonies *Bacillus amylobacter*.
    1 colony   Acid-fast bacillus.
    4 colonies stained by Gram.
    7   ,,   not stained by Gram.

Fungi :

    2 colonies of *Macrosporium* sp.
    3   ,,   *Penicillium glaucum*.
    1 colony   *Cladosporium herbarum*.
    1   ,,   *Fusarium roseum*.

C. Fly from refuse can.

Gave 116 colonies of bacteria and 10 colonies of fungi.

Bacteria:

34 colonies of *Bacillus coli commune.*
16   ,,   *Bacillus subtilis.*
 8   ,,   *Bacillus tumescens.*
 4   ,,   *Bacillus lactis acidi.*
12   ,,   *Sarcina lutea.*
 2   ,,   *Sarcina ventriculi.*
21   ,,   *Micrococcus pyogenes aureus.*
11   ,,   *Micrococcus ureae.*
 2   ,,   Acid fast bacilli.
 4 colonies stained by Gram.
 2   ,,   not stained by Gram.

Fungi:

4 colonies of *Penicillium glaucum.*
1 colony   *Eurotium* sp.
2 colonies   *Saccharomyces* sp.
1 colony   *Fusarium roseum.*
1   ,,   *Aspergillus niger.*
1   ,,   *Mucor racemosa.*

The large number and variety of bacilli carried by the fly *C* demonstrate in a very convincing manner the infection which a fly frequenting refuse is able to carry. All flies frequent refuse of varying degrees of filthiness and divide their attention between such refuse and other food, as is a matter of common observation. The significance of these observations therefore is of such a nature as to render unnecessary further discussion of the contention that all flies are germ-carriers.

CHAPTER VIII

THE DISSEMINATION OF TYPHOID FEVER BY FLIES
AND THEIR RELATION TO SUMMER DIARRHOEA

WHEN a Board of the highest qualified officers, who were appointed by the United States Government to investigate the extensive prevalence of typhoid fever in the military camps during the Spanish American war in 1898, when one-fifth of the soldiers contracted typhoid fever, refer to flies as 'These pests which had inflicted greater loss upon the American soldiers than the arms of Spain had,' many are compelled to readjust their opinions with regard to the character of the house-fly. Yet such is the case, and the comprehensive report of these officers, Drs Reed, Vaughan and Shakespeare, is one of the most striking indictments of the fly, indicating to what extent it can distribute bacterial organisms when the opportunity is offered. In those military camps, and again in the British camps in South Africa a few years later, ample opportunities for the carriage of infection by flies were certainly offered. Notwithstanding the instructions of Dr G. M. Sternberg, Surgeon-General of the United States Army, namely, that the surface of the faecal matter in the latrines or sinks should be covered with fresh earth, quick-lime or ashes three times a day, the latrines swarmed with flies. Veeder states that in the camps he saw faecal matter fresh from the bowel and in its most dangerous condition covered with myriads of flies, and at a short distance were the men's tents equally open to the air for dining and cooking. Others observed flies with their feet white with the lime from the latrines, on the food. What more perfect conditions could be required for the dissemination of the typhoid bacillus than these: open latrines frequented by incipient or convalescent cases of typhoid, millions of flies breeding in the latrines and visiting the mess tents a short distance away? Where the mess tents were protected from the flies with mosquito netting little or no typhoid was contracted. When the cold weather came the flies disappeared and with them the typhoid. The conclusion which the Board of investigation came to was: 'Flies undoubtedly served as carriers of infection.'

Dr G. M. Kober would appear to be the first who suggested the possible transference by flies of typhoid germs from infected faeces to food in a report on the prevalence of typhoid fever in the District of Columbia in 1895. But it was not until the exhaustive inquiry to which I have briefly referred was made that the public realised that the house-fly was something more than a mere irritating and disgusting nuisance but a

serious factor in the spread of one of our most prevalent infectious diseases and must be treated as such.

The significance of the danger will be appreciated
if certain of the features of typhoid fever are
considered. It is an intestinal disease and caused
by the entrance of the typhoid bacilli into the
digestive system by way of the food or drink. A
person may be distributing the typhoid bacilli in the
excreta for ten or more days before being laid
low by the fever, but still more serious than this is
the fact that not only does the convalescent typhoid
fever patient continue to excrete typhoid bacilli in a
number of cases but in certain cases the patient may
have, to all appearances, completely recovered and
yet continue excreting the typhoid bacilli. These
cases are known as 'carriers' and the number of
such 'carrier' cases who continue to carry the
typhoid germs in their bodies is gradually increasing
as the matter is being more carefully investigated.
It is being discovered that many of the inexplicable
outbreaks of typhoid fever are due to the presence of
one of the chronic carriers. The presence of an
unrecognised 'carrier,' excreting infected matter,
the occurrence of large numbers of flies and their
access to food or milk are all the factors that are
required to initiate an epidemic of typhoid fever, and
not a few epidemics are now being traced to the
concurrence of these factors.

In the military camps during the last Boer war in South Africa the conditions existing in the United States military camps were reproduced with the result that about 30 per cent. of the deaths were due to typhoid fever, flies proving as deadly as bullets. Almost without exception the surgeons who served in South Africa comment on the enormous numbers of flies frequenting the latrines and the mess tents. In some cases the typhoid patients in the hospitals could be distinguished from the other patients by the way the flies clustered round their mouths and eyes while in bed. There was a remarkable consensus of opinion as to the importance which flies played in the spread of infection when fever had been introduced into a camp. Similar experiences are related of camp conditions in other parts of the world such as India and the West Indies. In our own towns and cities the same conditions are frequently reproduced on a smaller scale. Flies may be found breeding in millions in heaps of stable refuse and the stables are frequently situated in insanitary sections of the town where old conservancy methods such as pails and privy middens are still used. Should a case of typhoid be introduced the disease not infrequently makes its appearance shortly afterwards in neighbouring houses, having been carried by flies from infected matter to food. It is a significant fact that wherever old conservancy methods are being replaced by the more sanitary system of water carriage thereby removing the possibility of flies becoming infected, there has been a decrease in the typhoid fever rate. This fact is well known to medical officers of health.

We have been considering what may be termed the circumstantial or epidemiological evidence of the carriage of typhoid fever by flies and while this is most conclusive the argument is clinched by the bacteriological evidence, the result of exact experiments. These experiments indicate that the fly is able to carry the typhoid bacillus in a viable condition either externally, that is on its body and appendages, or internally where its length of life is increased considerably. The typhoid bacillus is a non-spore-bearing bacillus which means that it is less adapted to external transference than a spore-bearing bacillus such as the anthrax bacillus. On this account, it is not improbable that the more usual method of infection is from the alimentary tract of the fly, either by the regurgitated vomit or by the faecal spots, as the bacilli will persist for a greater length of time in the alimentary tract than on the body or appendages. Ficker recovered the typhoid bacillus from flies twenty-three days after they had been infected by feeding on typhoid infected matter. Graham-Smith has shown that the bacillus may remain alive in the intestine of the fly for six days after feeding and that flies may infect surfaces upon which they walk for at least forty-eight hours after infection. In the case of *Bacillus enteritidis* he found that the flies could infect the surfaces over which they walk for some days and it would appear that the infection is largely due to inoculation by the flies' proboscides. This method of infection is of very great importance in the case of such foods as sugar and milk, as the same investigator has shown that flies infect both milk and syrup on which they feed or into which they fall. A number of investigators have recovered the typhoid bacilli from flies caught in the neighbourhood of cases of typhoid fever. Ficker recovered the bacillus from flies caught in a house in Leipzig where eight cases of typhoid had occurred. In Chicago, Hamilton recovered the bacillus five times in eighteen ex- periments from flies caught in two undrained privies, on the yard fences and walls of two houses and the room of a typhoid fever patient. The experiments of Faichne corroborate those of Graham-Smith as to the importance of the intestine of the fly as a carrier of the infected material. Faichne working in India was of the opinion that the fact that the flies had bred in infected matter was of greater importance as affecting their bacilli-carrying possibilities, than the fact that they had walked over or in other ways come in contact with infected matter. He therefore carried

6—2

out a careful series of experiments with a view to testing the validity of this idea. He reared the maggots in typhoid infected faeces and after the necessary precautions with regard to sterilisation, he was able to show that the flies which ultimately developed from such maggots contained virulent typhoid bacilli in their intestines. The results of Faichne's observations are of very great practical importance in their bearing on the control of typhoid fever in military camps or under civil conditions. In India Aldridge hatched 4042 flies (*Musca domestica* sub sp. *determinata*) from one-sixth of a cubic foot of soil taken from a trench. In conjunction with this the observations of Graham-Smith on the rate of defaecation and regurgitation of flies should be taken. He found that flies defaecated from three to eleven times per hour and vomited from six to fifteen times an hour, the rate depending upon the temperature and food. If these results are taken in conjunction with each other some idea will be gained of the manner in which and extent to which typhoid may be spread by flies which have been bred in infected excreta in camp latrines and have access to food.

In cities and towns, and also in country districts, the conditions and consequent dangers which have been described in military camps occur, though not usually to such an alarming degree. A single instance of many which are on record, may be mentioned as illustrating the way in which flies may disseminate infection. It occured in Denver, U.S.A. and is recorded by the Secretary of the Colorado State Board of Health. In August the wife of a dairyman was taken with typhoid fever and remained at home for about three weeks before being removed to the hospital at the end of August. In September numerous cases of fever were reported in the northern part of the city and on investigation it was found that all these cases had been obtaining their milk from this dairy. The conditions of the dairy were then inquired into and it was found that the dairy-man himself was suffering from a mild case of typhoid but was still delivering milk. The water supply was fairly good. It was found, however, that both the husband and wife had been using an open privy located within thirty-five feet of the milk-house which was unscreened and open to flies. A gelatin culture exposed for thirty minutes in the rear of the privy vault and in the milk-house among the milk cans gave numerous colonies of typhoid bacilli and colon bacilli. The source of infection of the dairy-man's wife's case was unknown. 'But,' the report states, 'I am positive that in all the cases that occurred on this milk route the infection was due to bacilli carried from the privy by the flies and deposited upon the milk cans, separator and utensils in the

milk-house, thereby contaminating the milk.' The
dairyman supplied milk to 143 customers. Fifty-five
cases of typhoid fever occurred and six deaths resulted
therefrom. This case speaks for itself and leaves
little more to be said with regard to the part which
flies may play in the dissemination of typhoid.

Summer or epidemic diarrhoea is one of the most
serious causes of the high infantile mortality, especially
in towns and cities. It is a rapidly spreading infectious disease and the evidence that it is conveyed
largely by flies is very strong. Nash was one of the
first to call attention in 1902, to the striking co-
incidence between the prevalence of this disease and
the number of flies. The summers of 1902–03 were
wet, consequently unfavourable to the breeding and
activity of house-flies, and these years were marked
by a low infantile mortality rate due to diarrhoeal
diseases being less prevalent. Medical officers of
health have repeatedly noted that where there are
large collections of refuse and other insanitary conditions serving as breeding places for flies, there is a
large number of cases of infantile diarrhoea. The
most careful and exhaustive study of the relation of
flies to infantile diarrhoea and its epidemiology has
been made by Niven (1910) who has carried on an
extensive study of this disease extending over a
number of years. He has made statistical observations on the numerical abundance of flies from week

to week during the fly's diarrhoea season and repeated these observations each year. As a result he found that the disease becomes more fatal only after the house-flies have been prevalent for some time and its fatality rises as their numbers increase and falls as they fall. The diarrhoea fatality corresponds more closely with the number of flies in circulation than with any other fact. The close correspondence between flies and cases of fatal diarrhoea receives a general support from the diarrhoeal history of sanitary sub-divisions in the Manchester district (see fig. 19). The methods of dissemination and the factors governing it are similar to those which have already been described in the case of typhoid fever. Niven observed specially that flies clustered about the nose and mouth of infants suffering from diarrhoea.

Up to the present time it has not been definitely proved what the causative organism of this infantile diarrhoea is. We are therefore compelled to rely almost wholly upon circumstantial or epidemiological evidence in discussing the disease. Morgan has isolated a definite bacillus which may be an important factor in the causation of the disease and this bacillus has been found in as many as 55·8 per cent. of cases of infantile diarrhoea examined. Rats and monkeys were infected with the bacillus by feeding and after a period succumbed to diarrhoea. It is also extremely interesting to note that Morgan's bacillus

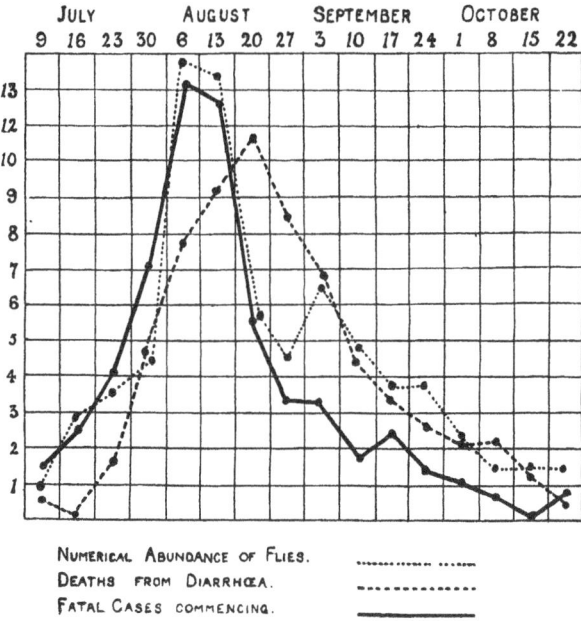

Fig. 19. Chart illustrating the relation of house-flies to summer
diarrhoea in 1904 in the city of Manchester. Prepared from
statistics and charts given by Niven (1910).

has been recovered from flies caught in infected and uninfected houses.

Hamer who has studied the relation of the abundance of flies to the number of cases of infantile diarrhoea has pointed out certain serious difficulties in the way of believing that flies are carriers of the causative organism or organisms of infantile diarrhoea. The chief of these is that at the end of the hot weather, while the number of flies is still very great, as was the case earlier in the season when the number of cases of diarrhoea was increasing, the number of cases begins to decrease. In explanation of this difficulty I have pointed out (1910) that the fall in the number of flies is usually preceded by a fall in the temperature which means a reduction in the activity of the flies, so that even though the flies were numerically abundant, their decreased activity would render them less liable to carry the causative organisms of summer diarrhoea. The number of cases of diarrhoea is dependent upon the activity and circulation of the flies. As this is a gradually increasing quantity at the beginning of the hot weather the number of cases of diarrhoea increases. When, at the end of the hot weather the activity of the flies is decreased, in spite of a numerical abundance of flies, the number of cases decreases. This apparent difficulty would rather support the idea of the flies' close relationship to diarrhoea than militate against it. Niven suggests that the more rapid diminution of the number of deaths may be due to the fact that the more susceptible and exposed infants have been killed off or rendered immune.

In the figure it will be observed that the rise and fall in the abundance of flies, which in this instance was measured by the numbers caught in twelve bell traps located in different parts of the city (Manchester), coincided very closely with the rise and fall in the curve representing the number of fatal cases of diarrhoea commencing, and anticipated remarkably closely, by about a fortnight, the curve representing the number of deaths from diarrhoea.

The circumstantial evidence in favour of flies playing an important *rôle* in the dissemination of infantile diarrhoea appears to me to be of so strong a character that one is justified in accepting it until its validity is disproved by subsequent investigation.

CHAPTER IX

THE RELATION OF FLIES TO CERTAIN OTHER INFECTIOUS DISEASES

ALTHOUGH the fly as a disease carrier is of importance chiefly in the dissemination of typhoid fever, for which reason the name 'typhoid fly' has been suggested by Howard, it is undoubtedly able to carry the organisms of certain other diseases when it

has access to infected material and therefore its relation to these diseases necessitates our attention.

Tuberculosis.

By the critical person the suggestion that the fly
may be concerned to any appreciable extent in the
dissemination of tubercle bacilli may not, at first
sight, be fully accepted, the means by which the
disease may be spread being numerous. A careful
examination of the facts, however, indicates that under
certain conditions they are able to play an important
rôle in the transfer of the *Bacillus tuberculosis*. The
partiality which flies display towards sputum is a
matter of common observation and their attentions to
unclean spittoons or cuspidors have not infrequently
filled one with disgust. These same flies having
infected themselves both externally and internally
also visit our food, the milk or cream jug, the infant's
feeding-bottle and mouth, and other substances and
articles upon which the bacilli can be deposited
either from the appendages or the body or by means
of vomit or faecal spots. In this last connection
Graham-Smith's observations are important. He
found that whereas flies fed on syrup produced an
average of 4·7 faecal deposits per day and those fed
on milk produced 8·3 deposits, the flies which fed on
sputum produced an average of 27 faecal deposits per

day and that the faeces were more voluminous and
liquid than usual. The possibility of infection therefore, in the case of such a bacillus as *B. tuberculosis*
occurring in sputum, would appear to be increased.

The fact that flies carry the tubercle bacilli has
been demonstrated for many years. Spillman and
Haushalter in 1887 found *B. tuberculosis* in the dejections and intestines of flies caught in a hospital
ward. Hoffman about the same time also found
virulent tubercle bacilli in the excreta of flies in a
room where a person had died of tuberculosis. In
1904 Hayward obtained tubercle bacilli in ten out of
sixteen cultures made from flies which had been
caught feeding upon bottles containing tuberculous
sputum, and virulent bacilli were obtained from these
faeces. In the same year Lord showed that flies may
ingest tubercular sputum and that the virulence of
the bacilli subsequently excreted may last for at least
fifteen days. Graham-Smith (1910) carried out a
number of experiments on flies artificially fed upon
B. tuberculosis. It was found that under experimental
conditions tubercle bacilli were present in the crop
for at least three days and in the intestine they
occurred in considerable numbers up to six days, being
still present twelve days after feeding. In the faeces
they were found to be numerous up to the fifth day
and they were occasionally found up to the fourteenth
day after infection. Flies which were fed first upon

tuberculous and afterwards upon non-tuberculous sputum were found to contain tubercle bacilli in their intestines for at least four days and their faeces were infected during the same period.

The experiments indicate most conclusively the manner in which flies may carry *B. tuberculosis* and that it is possible for them to distribute the bacilli for several days after feeding upon infected material, whether it be sputum or infected excreta from persons suffering from intestinal tuberculosis. From such infected matter they may carry the bacillus to food, and we now have a fuller knowledge of the danger of tubercular infection by way of the intestinal tract. Too much stress cannot be laid, therefore, upon the necessity of protecting tubercular sputum and matter from flies outside or inside the hospital, and the protection of food from flies in places where tubercular sputum is likely to be found is essential to the control of this disease.

Ophthalmia.

Most visitors to Egypt and to certain other tropical and sub-tropical countries have been struck not only by the extraordinary abundance of flies but by the manner in which they affect the eyes of the natives suffering from ophthalmia or conjunctivitis, an infectious inflammatory disease of the eyes. Similar

attention by flies to native Indians suffering from eye
trouble have been observed in Canada.   Under these
circumstances, therefore, it is not surprising that
earlier investigators recognised the part which flies
play in the dissemination of these inflammatory
diseases of the eyes.   Flies are always attracted to
exposed moist or inflammatory places on the body.
As early as 1862 Budd considered that flies undoubtedly acted as carriers of Egyptian ophthalmia.
Later, Laveran in 1880 stated that in the hot season
at Biskra the eyes of the native children were covered
with flies which would carry the infectious discharge
on their legs and proboscides to healthy children.
Eight years later Howe pointed out that the number
of cases of ophthalmia increases with the abundance
of flies and that the prevalence of the disease was
proportionate to the abundance of flies.   Although I
cannot find much evidence of a bacteriological nature
to support this idea, which is so conclusively sub-
stantiated by the circumstantial or epidemiological
evidence, it is interesting to note that Howe states
that an examination of flies captured on diseased
eyes revealed bacteria similar to those found in the
secretions produced by ophthalmia.   Flies which I
have had sent to me from Egypt proved to be *Musca
domestica*, and correspondence which I have had with
Dr Andrew Balfour of Khartoum and others in Egypt
supports the view as to the relationship of flies to

ophthalmia, and there are a number of investigators who have adduced evidence which is very conclusive.

Anthrax.

This disease of cattle, sheep and other animals is caused by a spore-bearing bacillus whose entry into the human system gives rise to ulcers generally known as 'malignant pustule.' The belief that anthrax might result from the bite of a fly was entertained as early as the eighteenth century. In these cases, however, reference was undoubtedly made to biting flies and while the mistaken idea of the house-fly's ability to bite was as prevalent then as it is at the present time, it is only under certain conditions that house-flies and their non-biting allies the blow-flies or 'blue-bottles' would be able to carry the anthrax bacillus. One not infrequently finds cases where the bite of the fly is responsible for an ulcerated or inflammatory wound. In such a case a number of biting flies might be responsible. It is conceivable that such flies as the Horse flies (*Tabanidae*) or their blood-sucking relatives the Golden-eyed *Chrysopidae* and Haematopota, which had been feeding upon diseased animals or upon putrefying matter, might introduce pathogenic bacteria when piercing the skin of man and give rise thereby to an ulcerated condition.

Under similar conditions the Stable-fly (*Stomoxys calcitrans*) might also cause trouble. For the house-fly or blow-fly, however, to transfer contagious matter it would be necessary for them to have access to a wound, or to infect food. In certain cases, for example in tanneries and places where the carcases of animals are handled, the first of these conditions might be conceivably fulfilled. In 1869 Raimbert proved experimentally that the house-fly and meat-fly were able to transfer anthrax bacilli on their proboscides and legs. Davaine confirmed these results in the following year in the case of the blow-fly (*Calliphora vomitoria*), and in 1874 Bollinger found the bacilli in the alimentary tract of flies that had been caught in the carcase of a cow which had died of anthrax. Passing over the experiments of other observers who merely allowed flies to walk over infected material and after-wards over culture plates upon which colonies of anthrax bacilli naturally grew as one would expect *à priori*, we come to the careful experiments of Graham-Smith (1910–11). The anthrax bacillus was found in the vomit of flies which had fed upon infected material in the shape of the body of a mouse which had died of anthrax. From further experiments it was shown that the bacillus did not remain in a virulent condition on the appendages of the fly for more than twenty-four hours, but they remained alive in the crop containing coagulated blood for five days. The bacilli were also found in the fly's faeces forty-eight hours after infection. The anthrax bacillus is a spore-forming bacillus and it is able in the spore stage to remain in a virulent condition for some length of time. Experimenting with flies fed upon anthrax spores, Graham-Smith found that this bacillus remained in a virulent condition for at least twenty days on the appendages and in the alimentary tract, and that faeces passed fourteen days after infection contained living spores. Dried faeces and vomit were proved to contain virulent spores for twenty days. Finally, it was demonstrated by inoculations that dead flies retained the bacillus in a virulent condition for four hundred and twenty-eight days.

In another series of experiments it was shown that blow-flies bred from larvae which had been allowed to feed on meat infected with anthrax spores were infected and capable of transferring the infection for two days after emerging from the pupal state. This last experiment is of practical importance as it indicates the necessity of preventing flies from having access to the carcases of animals which have died from anthrax either for the purpose of depositing their eggs or for feeding.

It will be seen therefore that under certain conditions which are not infrequently fulfilled, flies would disseminate the anthrax bacillus.

Cholera.

Moore in 1853 pointed out the necessity of protecting food from flies in the belief that they might
transfer cholera from infected excreta or other
materials; and the indiscriminate manner in which
flies visited in great numbers the dejections and food
of cholera patients was observed by Nicholas in 1849
and recorded by him in 1873. The latter observer
also noted the fact that abundance of flies was
synchronous with the prevalence of the disease, and
his observation was confirmed by later writers. A
number of earlier investigators among whom were
Maddox in 1892, Tizzoni and Cattani in 1886 and
Sawtchenko in 1892, demonstrated experimentally that
flies were able to transfer the cholera spirillum. The
last observer and also Simmonds found that flies were
able to carry the cholera spirillum in their intestines.
It had been found by Uffelman that not only can the
flies carry a large amount of infection, as many as
10,500 colonies having been obtained from a single
fly, but that they will infect food such as milk upon
which they have fed. The method of disseminating
cholera, therefore, is very similar to the dissemination
of typhoid fever. The fly is infected with the cholera
spirillum both internally and externally and is ac-
cordingly able to infect such fluids and substances
upon which it feeds. Ganon has shown that flies are

able to transmit infection twenty-four hours after infection. The practical bearing of these facts is forcibly illustrated by some observations made by Macrae in India in 1894. Boiled milk was exposed in different parts of the gaol at Gaya where cholera was present and where flies occurred in enormous numbers. This milk and milk exposed in the cowsheds became infected, thus indicating one of the ways infection was carried. Other observers in India have confirmed these facts. Speaking of a cholera outbreak in Northern China in 1902 a Japanese Army Surgeon called attention to the important part which flies must play in the spread of the disease, and his convictions were strengthened by his isolating the cholera spirillum from flies caught in an infected house in Tientsin.

The evidence both epidemiological and bacteriological that flies play an important part in the spread of cholera injection is of a strikingly convincing nature.

Plague.

The discoveries of recent years of the relation of fleas to plague have tended to minimise the idea which has been prevalent since the fifteenth century that flies bear some relation to plague and act as carriers of the infection; and while it is no doubt true that as an epidemiological factor they are not of so great consequence in this disease as in certain of the diseases which have already been discussed, the experiments of Nuttall and others indicate that flies should not be allowed to have access to the bodies or excreta of plague patients or to food. Nuttall demonstrated that house-flies which had been fed upon infected material might survive for at least eight days after feeding and that for forty-eight hours they carried the plague bacillus in a virulent state.

Other Diseases.

There are several other contagious diseases among which are Yaws and Tropical Sore, which are of such a nature as to permit the causative organisms to be carried by flies, and experiments have confirmed the beliefs that observers have held as a result of circumstantial evidence. The countless numbers of flies which affect natives, peculiarly indifferent to their attentions when they are suffering from these ulcerous affections, are a constant source of infection.

CHAPTER X

HOUSE-FLIES IN RELATION TO (1) MYASIS OF THE
INTESTINAL AND URINARY TRACTS AND (2) THE
SPREAD OF PARASITIC WORMS

(1) *Myasis of the intestinal and urinary tracts.*

MYASIS is the term applied to a diseased condition resulting from the occurrence of the larvae of flies of different species in the intestinal and urinary tracts. Scattered through medical and entomological literature are numerous cases of the occurrence of 'maggots' in the human intestine from which they have been expelled either by vomiting or by diarrhoeal affections. These 'maggots' have not always been carefully examined with a view to determining the species of fly to which they belong. The species of flies whose larvae are most generally responsible for these troubles of the intestines and the urinary tracts are the house-fly, *Musca domestica*, the Lesser house-fly, *Fannia canicularis*, the Latrine-fly, *Fannia scalaris* and the blow-fly, *Calliphora erythrocephala.*

When the larvae are present in the stomach they may cause violent pains and dizziness and are often expelled by vomiting. In the intestine they give rise to abdominal pains and diarrhoea, not infrequently accompanied by haemorrhage caused by the larvae

perforating the mucous lining of the intestine. Cases
of the occurrence of larvae in the intestine have been
recorded since 1809. In some instances the larvae
appeared to have been present in the digestive tract
for a considerable length of time. In a case recorded
by Jenyns in 1839 the symptoms were noticed in
spring but the larvae were not evacuated until the
summer or autumn following.

The occurrence of fly larvae in the urinary tract
is more remarkable than these occurrences in the
intestine, nevertheless we have records of a number
of cases occurring since the first apparent record in
the seventeenth century. Expulsion of these larvae
from the female urinary tract is more readily understood than their expulsion with the urine of a man.
The habits of the fly and the larvae render several
methods of infection possible. All the aforementioned
species of flies are attracted to decaying animal or
vegetable products, excrementous or purulent substances, for the purpose of depositing their eggs or
feeding. Should flies deposit their eggs upon decaying
fruit and other food the eggs or young larvae would
be thus taken into the digestive tract. Intestinal infection may result from the flies, which frequent privies
for the purpose of ovipositing, depositing their eggs in
the anal region, and the larvae on hatching enter by
way of the rectum. This mode of infection may be
a common one in the case of careless mothers who

leave their infants in an exposed and unclean condition.

The infection of the urinary ducts, however, is more difficult to understand, especially in the case of the male. Flies would be attracted to the genital apertures by the various albuminous and other secretions especially in cases where the organ is exposed for any length of time. During the hot weather the infection of both sexes by *Fannia canicularis* which frequents bedrooms might be possible.

(2) *The spread of parasitic worms by flies.*

The dissemination of many species of parasitic worms which infest the human intestine is rendered probable by the fact that the eggs of the worms pass out with the excreta, and this, especially when it is fresh, is attractive to flies which feed greedily upon the moist surface and also deposit their eggs upon it. Although the eggs of these parasites are comparatively large, measuring from ·01 mm. to ·15 mm. in length, the transference of certain of them upon the legs of the fly or inside its body might be expected. These beliefs have been confirmed by a number of observers.

Grassi in 1883 broke up segments of the common tapeworm *Taenia solium* in water. Flies sucked up the eggs in the water and they were found in the faeces of flies unaltered. The eggs of another species

of worm were also passed unaltered. The eggs of a
third species of parasitic worm *Trichocephalus* were
found in the faeces deposited in the room below the
laboratory and flies caught in the kitchen were found
to have their intestines full of eggs. Calandrucio
in 1906 showed that the eggs were infective after
passing through the intestines of the fly, and also
indicated a method by which infection might take
place. A number of flies were fed on infective
material containing eggs of a tapeworm (*Hymenolepis
nana*), common in Southern Italy, and they afterwards defaecated upon sugar. The eggs were found
in the intestines of some of the flies, and by means
of the infected and fouled sugar a girl was infected
and the eggs of this worm were found in her stools
twenty-seven days later. Another observer Galli-
Valerio found that flies could carry not only the eggs
but also the larvae of the hook-worm, one of the
most dangerous of parasitic worms affecting man.
Stiles found that flies bred from larvae which had
fed upon parasitic worms showed both eggs and
larval parasitic worms in their intestines.

A long series of experiments have been carried
out more recently by Nicoll (1911). He observed
that flies appeared to select the tapeworms in
preference to faeces where the two were given them
in a mixed condition, which indicated how readily
they may become infected externally or internally

with the parasitic larvae or eggs. Fly larvae would devour parasitic worms with great rapidity, a few larvae devouring a worm 20 to 30 times their own bulk within two or three days. Nicoll found that the house-fly was able to suck the eggs out of the tapeworm *Diplydium caninum* and carry them for at least 43 hours in its intestine. Both *M. domestica* and *F. canicularis* were able to ingest eggs of *Taenia marginata* which are comparatively large as they measure ·035 mm. × ·0355 mm. in size. These eggs were found in the intestines of the flies up to the third day after feeding. It was also shown that the house-fly *Musca domestica* can easily ingest the eggs of the tapeworm *Taenia serrata* both from faeces and from unbroken segments of the worm. Faeces containing tapeworm segments may continue to act as sources of infection, from which flies could infect food such as sugar, for as long as fourteen days. Nicoll was not able to find eggs of parasitic worms in flies bred from larvae which had been allowed to breed in faeces containing female parasitic worms. In a very interesting experiment it was shown that in spite of flies cleaning themselves, as they are always endeavouring to do, and notwithstanding their transference three times into clean glass vessels, they still carried on their legs and bodies eggs of the tapeworm *Hymenolepis diminuta*, upon which, mixed with faeces, the flies had been allowed to feed.

Flies carrying eggs on their legs and bodies were also found to infect sugar.

The experiments of these investigators, coupled with the feeding habits of the fly, indicate that there is a very strong probability that flies play a not unimportant part in the dissemination of parasitic worms affecting man.

CHAPTER XI

PREVENTIVE AND CONTROL MEASURES

' Within the past few years the knowledge of the causes of disease has become so far advanced that it is a matter of practical certainty that by the unstinted application of known methods of investigation and consequent controlling action, all epidemic diseases could be abolished within a period so short as fifty years.'

SIR RAY LANKESTER.

As it has been proved that the house-fly plays an important part in the dissemination of certain of our most prevalent infectious diseases, when the necessary conditions are present, its control becomes a necessary factor in any system of preventive medicine or sanitary reform. Its disgusting and irritating habits would afford sufficient reason for adopting preventive measures were it not known to be a carrier of disease germs. It is no exaggeration to say that in temperate climates it is as important a disease carrier as the

mosquito is in warmer countries, and the danger of
its disseminating disease increases with an increase
in temperature either in time or space.

The measures for the control and prevention of
the fly danger must have two objects :
(1) to prevent the flies breeding,
(2) to prevent the transference of infection by
them.

The first of these objects is fundamental and of
prime importance. Until this is realised any steps
which may be taken will be of no value. The
problem is similar in all respects to the mosquito
problem. The abolition of the breeding places of the
mosquito and the prevention of their breeding are
the first steps in the control of this insect and the
prevention of malarial fever. So also with the
house-fly.

The study of the breeding habits of the house-fly
has indicated the places and materials in which it
breeds. The chief breeding places are collections of
horse manure or stable refuse and it has been shown
previously in what enormous numbers flies breed in
this substance. The first steps, therefore, must be
taken in the direction of preventing flies from breeding
in horse manure or stable refuse.

It is obvious that the stable manure must not be
stored in places accessible to flies. Fly-proof receptacles or chambers should be provided in which the

manure is immediately placed. Recognising the importance of this many cities in the United States and Canada have passed bye-laws requiring that where horses, cows or similar animals are kept, properly constructed pits or chambers shall be maintained for the temporary storage of the manure. Such receptacles should be of solid masonry or concrete and provided with doors that will prevent the ingress of flies. Further, regulations governing the building of stables and cowsheds should provide that the floors of such places should be solid masonry or concrete. In conjunction with proper storage of the manure, its periodic removal must be provided for. With the greatest care it is not always possible absolutely to prevent flies from depositing their eggs in the manure. It has frequently been observed that the flies will deposit their eggs in the excreta immediately they are dropped, in fact, they prefer the warm excreta for this purpose. Therefore, to prevent the emergence of flies from eggs so deposited and stored with the refuse, the manure must be removed well within the shortest time that is taken for the life-cycle of the fly to be completed. In summer this time is shorter than in winter. Manure should be regularly removed at intervals not exceeding seven days during the summer and autumn months from June to October and at intervals of not more than nine days during the remainder of the year. The removal should be to

a place well without the range of flight of flies from
the nearest human habitations. The prohibition of
the storage of stable refuse, etc., in places such as
railway and other depôts pending its removal cannot
be insisted upon too strongly, as it has been found
that such practices result in an unusual and dangerous
abundance of flies in the neighbourhood of such
places for storage.

Whenever possible, in stables and in similar
places, in addition to the careful storage of the
manure in suitable receptacles, it should be treated
with an insecticidal substance which would kill any
larvae which might be in the manure and prevent
the emergence of flies. Many experiments have been
carried out by Howard, Herms and others with a view
to finding an insecticide which will be effective and at
the same time reasonably practicable. It has been
found that if a barrel of chloride of lime be placed in
the stable at the door of the manure pit or chamber
and a small shovelful of the lime be scattered over
the fresh manure after it has been thrown into the pit
the breeding of the flies is effectively controlled.
This plan was tried by one of my correspondents who
owned a stable of 150 horses. Not only was the lime
lightly scattered over the manure, but also over the
floors of the stables when they were cleaned out. He
praised the use of the lime not only on account of
its insecticidal value, but also because the horses

were able to rest instead of being incessantly tormented by flies, with the result that they were more fit for work. The last fact can be readily appreciated and should appeal to those who keep horses. It is doubtful whether the admixture of a small amount of chloride of lime would seriously affect the manurial properties of the stable manure.

Forbes, experimenting in Illinois, U.S.A., found that a solution of iron sulphate sprayed over the manure effectively controlled the breeding of the house-fly in horse manure. A solution of two pounds of iron sulphate in one gallon of water for each horse per day was used. It was calculated that the average city horse produces about fifteen pounds of manure per day and the heavier draught horses produce twenty to thirty pounds per day. The amount to be treated, however, is less than this, as the horses are out of the stables for a large proportion of the day. Not only does the iron sulphate kill the larvae but it also deodorizes the manure and does not injure its manurial properties.

The old-fashioned and insanitary privy is not only a favourite breeding place but the commonest source of infection. Medical officers of health are unanimous in their condemnation of these mediaeval survivals, and their relation to infantile mortality is strikingly shown in the statements and evidence collected by the Medical Officer of Health of the Local Government

Board (1910), to which the reader desiring further information on these and other factors of infantile mortality is referred. In his general summary Dr Newsholme states: "Infant mortality is highest in those counties where, under urban conditions of life, filthy privies are permitted, where scavenging is neglected and where streets and yards are to a large extent not 'made up' or paved." In his recommendations he says: "Sanitary authorities in compactly populated districts should decide to remove all dry closets if a water-carriage system is practicable."

The danger to which helpless infants are exposed in our populous districts will be apparent to anyone who will give this matter a moment's consideration. Breeding in the excreta in their thousands, and by their emergence and frequent visits to their birth places befouling their bodies and limbs and drenching their intestines with whatever germs the filth may contain, the flies swarm over the faces, the food and the feeding bottles of the helpless infants. Under such circumstances it would be surprising if insanitary conditions did not bear a close relation to infantile mortality. This, however, has been more fully discussed in a previous chapter and we are concerned here with the prevention of the breeding of flies. The abolition of the insanitary privy is a necessary step in the control of the house-fly and the improvement of sanitary conditions. Such abolition is invariably accompanied by a reduction of the death rate due to intestinal disease and there can be little doubt that the destruction of the breeding places of the fly and its sources of infection which such sanitary improvements effect plays an important part in bringing about this reduction. In view of the incontrovertible evidence of the relationship of such insanitary conditions to infantile and other mortality, a serious responsibility is attached to local authorities permitting the existence of such conditions.

The third important breeding place includes what may be collectively termed 'organic refuse.' A list of the chief of these substances has been given already in describing the breeding habits of the fly. Such organic substances are frequently found in domestic refuse receptacles or heaps and in public tips or dumps. Wherever such collections are maintained flies will breed and infect themselves with putrefactive and other bacteria. The keeping of organic refuse in fly-proof receptacles and its prompt destruction are the only satisfactory means of depriving flies of such breeding places.

Many Health Departments have promulgated ordinances for the purpose of abolishing the fly nuisance as a means of disease prevention. The orders issued in 1906 by the District of Columbia, U.S.A., afford an excellent example of the orders which health authorities should have enforced. These have been summarised by Howard as follows:

'All stables in which animals are kept shall have the surface of the ground covered with a water-tight floor. Every person occupying a building where domestic animals are kept shall maintain in connection therewith a bin or pit for the reception of the manure, and, pending the removal from the premises of the manure from the animal or animals, shall place such manure in said bin or pit. This bin shall be so constructed as to exclude rain water, and shall in all other respects be water-tight except as it may be connected with the public sewer. It shall be provided with a suitable cover and constructed so as to prevent the ingress and egress of flies. No person owning a stable shall keep any manure or permit any manure to be kept in or upon any portion of the premises other than in the bin or pit described, nor shall he allow any such bin or pit to be over filled or needlessly uncovered. Horse manure may be kept tightly rammed into well covered barrels for the purpose of removal in such barrels. Every person keeping manure in the more densely populated parts of the District shall cause all such manure to be removed from the premises at least twice every week between June 1 and October 31, and at least once every week between November 1 and May 31 of the following year. No person shall remove or transport any

manure over any public highway in any of the more
densely populated parts of the District except in a
tight vehicle which, if not enclosed, must be effectually
covered with canvas, so as to prevent the manure
from being dropped. No person shall deposit manure
removed from the bins or pits within any of the more
densely populated parts of the District without a
permit from the Health Officer.'   As a further means
of preventing the spread of infection certain departments of health require the protection of foodstuffs
and milk sold or offered for sale and also the proper
storage or disposal of organic refuse or garbage.

The entire matter of the prevention of the breeding
of flies and the attitude which local health authorities
should take towards the question may be summarised
as follows: It has been proved by incontrovertible
evidence that house-flies are able to act as carriers of
the germs of certain prevalent infectious diseases,
and that their habits render them specially adapted
to the dissemination of disease. The presence of flies,
therefore, is a serious menace to the public health.
Further, it has been shown that flies breed in stable
refuse, insanitary privies and different forms of organic
refuse.   Therefore, to maintain such breeding places
constitutes a public nuisance of a very grave nature.
Most health authorities have powers to compel the
abatement of public nuisances and in view of the
foregoing facts such powers might be exercised

justifiably for the purpose of abolishing the breeding places of flies.

For the killing of flies in houses several remedies have been recommended such as burning of pyrethrum powder, the evaporation of carbolic acid on a hot iron shovel and poisoning with a dilute solution of formalin. R. J. Smith of the North Carolina Experiment Station, U.S.A., has shown that if formalin is mixed with sweet milk it proves very attractive to flies and the solution makes an excellent and fatal bait. One ounce or two tablespoonfuls of forty per cent. formalin is mixed with sixteen ounces, that is, one pint, of equal parts of milk and water. If this mixture is exposed in shallow plates in the middle of each of which a piece of bread is placed for the flies to alight upon, the flies will be attracted to the solution and poisoned. The formalin has also the advantage of being a disinfectant.

In view of what has already been stated concerning the different ways in which flies transfer infection, the means to be adopted to prevent such transference will be apparent. Briefly these means are, first, the protection of infected matter from flies, and second the protection of food both liquid and solid and the protection of the faces of infants and invalids from flies. The necessity of preventing flies from gaining access to excreta, infected or non-infected, is too obvious to need insisting upon, nor should flies have access to tubercular sputum or purulent discharges. The screening of food, of hospitals, of the sick room, and of infants is a measure which should be adopted as a matter of course rather than a hygienic necessity.

It is safe to say that if measures were taken to prevent flies breeding by the abolition or proper care and protection of possible breeding places on the one hand, and on the other hand to prevent them transferring infection from infected material, that the house-fly would cease to be a serious factor in the carriage of typhoid fever, tuberculosis and intestinal diseases of infants.

Many inquirers have asked whether house-flies do not perform some good service, in accordance with the doctrine that there is some good purpose bound up in the existence of every creature. To such persons my reply is in the affirmative. House-flies certainly perform a valuable service; they indicate the presence of filth and are the sanitarian's danger signals, his red lamps in fact. House-flies are indications of the fact that insanitary conditions are present, that the machinery requisite for an epidemic of typhoid fever is in excellent working order and that more children are dying each year from intestinal disease than should be the case. When these facts are realised, the house-fly will stand out in its true light, as a potential destroyer of human life.

BIBLIOGRAPHY

The following list contains only the chief memoirs referred to in the foregoing chapters. Those who may desire further references should consult the bibliographies contained in the memoirs published by the present author in *The Quarterly Journal of Microscopical Science* (see below), the bibliographical list contained in Dr Howard's popular account (Howard 1911) and the summary compiled by Dr Nuttall and Mr Jepson (1909).

1911. COPEMAN, S. M., HOWLETT, B. A., and MERRIMAN, G. 'An experimental investigation on the range of Flight of Flies.' *Reports to the Local Government Board on Public Health and Medical Subjects*, N. S., No. 53, pp. 1–9.

1790. VON GLEICHEN, F. WILHELM. 'Geschichte der gemeinen Stubenfliege.' 32 pp. 4 pls. ; Nürnberg.

1909. GRAHAM-SMITH, G. S. 'Preliminary Note on Examinations of Flies for the presence of colon bacilli.' *Reports of the Local Government Board on Public Health and Medical Subjects*, N. S., No. 116, pp. 9–13.

1910. —— 'Observations on the ways in which artificially infected flies (*Musca domestica*) carry and distribute pathogenic and other bacteria.' *T. c.*, N. S., No. 40, pp. 1–41 ; 7 pls.

1911. —— 'Further observations on the ways in which artificially infected flies carry and distribute pathogenic and other bacteria.' *T. c.*, N. S., No. 53, pp. 31–48.

HOUSE-FLIES

1907. HEWITT, C. G. 'The Structure, Development, and Bionomics of the House-fly, *Musca domestica* Linn.: Part I.
The Anatomy of the fly.' *Quart. Journ. Micr. Sci.*, Vol.
51, pp. 395–448, pls. 22–26.

1908. —— Idem, Part II. 'The Breeding Habits, Development
and the Anatomy of the Larva.' *Ibid.*, Vol. 52, pp. 495–
545, 4 pls.

1909. —— Idem, Part III. 'The Bionomics, Allies, Parasites,
and the Relations of *M. domestica* to Human Disease.'
Ibid., Vol. 54, pp. 347–414, 1 fig., 1 pl.

1908. —— 'The Biology of House-flies in Relation to the
Public Health.' *Journ. Roy. Inst. Public Health*, Vol. 16,
pp. 596–608, 3 figs.

1910. —— 'The House-fly. A Study of its Structure, Development, Bionomics and Economy.' Manchester University
Press. Biological series, No. 1. xiii + 195 pp., 10 pls.
(This is a republication in volume form in a limited edition
of 200 copies of the reprints of the three memoirs published
in the *Quart. Journ. Micr. Science*, 1907, 1908 and 1909.
It is now out of print and a revised account is being
prepared.)

1900. HOWARD, L. O. 'A contribution to the study of the insect
fauna of human excrement (with especial reference to the
spread of Typhoid fever).' *Proc. Wash. Acad. Sci.*, Vol. 2,
pp. 541–604, figs. 17–38, pls. 30–31.

1911. —— 'House-flies.' *Farmers Bulletin*, No. 459. *U. S.
Dept. Agr.* pp. 16, and figs.

1911. —— 'The House fly: Disease carrier: An account of
its dangerous activities and of the means of destroying it.'
xix + 312 pp. 38 figs. New York.

1910. NEWSHOLME, A. 'A report on infant and child mortality,
being a Supplement to the Report of the Medical Health
officer in the 39th Annual Report of the Local Government Board 1909–10.' (Separate, 110 pp.)

1911. NICOLL, J. W. 'On the part played by flies in the dispersal of the eggs of parasitic worms.' *Reports to the Local Government Board on Public Health and Medical Subjects*, N. S., No. 53, pp. 13–30.

1910. NIVEN, J. 'The House-fly in Relation to Summer Diarrhoea and Enteric Fever.' *Proc. Roy. Soc. of Medicine.* April, 1910. (Reprint, 83 pp.)

1909. NUTTALL, G. H. F., and JEPSON, F. P. 'The part played by *Musca domestica* and allied (non-biting) flies in the spread of infective diseases. A summary of our present knowledge.' *Reports of the Local Government Board on Public Health and Medical Subjects*, N. S., No. 16, pp. 13–41.

1911. RANSOM, B. H. 'The life-history of a parasitic nematode (*Habronema muscae*).' *Science*, N. S., Vol. 34, pp. 690–692.

INDEX